CQR Pocket Guide
Algebra II

Helpful Formulas

Following are some formulas that you are likely to need as you delve into Algebra II. To aid you in your studies, you may want to memorize these for those times when you can't (or don't) have this sheet with you.

Geometry Formulas

To Find	Use this Formula
Perimeter	$P = $ sum of sides
Area of triangle	$A = \frac{1}{2}bh$
Area of a rectangle	$A = bh$
Area of a parallelogram	$A = bh$
Area of a trapezoid	$A = \frac{1}{2}h(b_1 + b_2)$
Area of a circle	$A = \pi r^2$

Word Problem Formulas

To Find	Use this Formula
Distance	$Distance = rate \times time$
Interest	$Interest = principal \times rate \times time$
Compound Interest	(compound) $I = p + prt$
Work accomplished	work $accomplished = (rate$ of work$)$ $\times (time$ worked$)$

Graphing Formulas

To Find	Use This Formula
Distance (for graphs)	$\sqrt{(x_2 - x_1)^2 + (y_2 - y_1)^2}$
Midpoint	$M = \left(\frac{x_1 + x_2}{2}, \frac{y_1 + y_2}{2} \right)$
Slope	$m = \frac{y_2 - y_1}{x_2 - x_1}$ Note: $x_2 \neq x_1$
Slope and y-intercept	$y = mx + b$ (**Note:** m is slope; b, c, and d are constants)
Line parallel to above	$y = mx + c$
Line perpendicular to both	$y = -\frac{1}{m}x + d$

CQR Pocket Guide
Algebra II

Working Algebraic Fractions

Algebraic fractions are worked with in the same way as common fractions.

- To add or subtract, use common denominators.
- Multiply numerators together and denominators together.
- Division is reciprocal multiplication. Cross products work only as follows:

$$\frac{a}{x} = \frac{b}{y} \;\rightarrow\; ay = bx$$

Quadratic Equations

Standard form $\quad ax^2 + bx + c = 0$

Difference of squares $\quad x^2 - b^2 = (x + b)(x - b)$

Quadratic formula $\quad x = \dfrac{-b \pm \sqrt{b^2 - 4ac}}{2a}$

Sequencing

To find the nth term of a sequence, where d is the common difference, and r is the ratio of consecutive terms:

Arithmetic $\quad a_n = a_1 + (n - 1)d$

Geometric $\quad a_n = a_1 r^{n-1}$

Relations

Memorize the following two relations:

$$\sqrt{-4} = 2\sqrt{-1} = 2i \qquad\qquad x = 2^y \equiv y = \log_2 x$$

Conic Sections

Circle [centered at (h,k)] $\quad (x - h)^2 + (y - k)^2 = r^2$

Ellipse $\dfrac{x^2}{a^2} + \dfrac{y^2}{b^2} = 1 \qquad (a^2 > b^2)$

Hyperbola $\quad \dfrac{x^2}{a^2} - \dfrac{y^2}{b^2} = 1$

Parabola $\quad y = a(x - h)^2 + k \;$ or $\; x = a(y - k)^2 + h$

For more information about Wiley, call 1-800-762-2974.

CliffsQuickReview™
Algebra II

By Edward Kohn, MS and David Alan Herzog

WILEY

Wiley Publishing, Inc.

About the Authors

Ed Kohn is an outstanding educator and author, with over 33 years' experience teaching mathematics. Currently, he is the testing coordinator and math department chairman at the Sherman Oaks Center for Enriched Studies. Formerly an instructor at Fairleigh Dickinson University, David Alan Herzog is the author of over 100 audio visual titles and computer programs and has written and edited several books in mathematics and science.

Publisher's Acknowledgments

Editorial

Project Editor: Tracy Barr

Acquisitions Editor: Sherry Gomoll

Technical Editor: Mike Kelley

Editorial Assistant: Michelle Hacker

Composition

Indexer: TechBooks

Proofreader: Christine Pingleton

Wiley Indianapolis Composition Services

CliffsQuickReview™ *Statistics*

Published by
Wiley Publishing, Inc.
909 Third Avenue
New York, NY 10022
www.wiley.com

Copyright © 2001 Wiley Publishing, Inc. New York, New York

Library of Congress Control Number: 2001024141

ISBN: 0-7645-6371-8

Printed in the United States of America

15 14 13 12

1O/RV/QX/QU/IN

Table of Contents

INTRODUCTION

CliffsQuickReview *Algebra II* is a comprehensive study guide to the many topics of a second course in algebra, including information on subjects ranging from linear equations in one, two, and three variables, through inequalities, complex numbers, conic sections, quadratic equations, and logarithms. Whether you are looking for an in-depth treatment of the entire subject matter of the course or occasional reinforcement of one or more aspects of algebra, this is the place to find it.

For the purpose of this review, your knowledge of the following fundamental ideas is assumed:

- Natural numbers

- Whole numbers

- Integers

- Rational numbers

- Irrational numbers

- Real numbers

- Rules for positive and negative numbers

- Exponents

- Grouping symbols

- Order of operations

- Scientific notation

- Absolute value

If you feel that you are weak in any of these topics, refer to CliffsQuickReview *Algebra I*.

Why You Need This Book

Can you answer yes to any of these questions?

- Do you need to review the fundamentals of Algebra II fast?

- Do you need a course supplement to Algebra II?

- Do you need to prepare for your Algebra II test?

- Do you need a concise, comprehensive reference for Algebra II?

- Do you need a no-nonsense approach to Algebra II that gets you the results you need?

If so, then CliffsQuickReview *Algebra II* is for you!

How to Use This Book

You're in charge here. You get to decide how to use this book. You can read it straight through (not usually a good idea with a math book) or just look for the information that you want and then put the book back on the shelf for later use. Here are a few of the recommended ways to search for information about a particular topic:

- Use the Pocket Guide to find essential information, such as important algebraic formulas and concepts.

- Look for areas of interest in the book's Table of Contents or use the index to find specific topics.

- Flip through the book, looking for subject areas at the top of each page.

- Get a glimpse of what you'll gain from a chapter by reading through the "Chapter Check-In" at the beginning of each chapter.

- Use the "Chapter Checkout" at the end of each chapter to gauge your grasp of the important information you need to know.

- Test your knowledge more completely in the CQR Review and look for additional sources of information in the CQR Resource Center.

- Look in the Glossary for important terms and definitions. If a word is boldfaced in the text, you can find a more complete definition in the book's glossary.

Visit Our Web Site

A great resource, www.cliffsnotes.com, features review materials, valuable Internet links, quizzes, and more to enhance your learning. The site also features timely articles and tips, plus downloadable versions of many CliffsNotes books.

When you stop by our site, don't hesitate to share your thoughts about this book or any Wiley product. Just click the Talk to Us button. We welcome your feedback!

Chapter 1

LINEAR SENTENCES IN ONE VARIABLE

Chapter Check-In

❏ Solving linear equations

❏ Solving problems by using formulas

❏ Understanding linear inequalities

❏ Dealing with compound inequalities

❏ Solving absolute value inequalities

Linear sentences may be equations or inequalities. What they have in common is that the variable has an exponent of 1, which is understood and so never written. They also can be represented on a graph in the form of a straight line (covered in Chapters 2 and 3). This chapter examines many ways of writing and solving linear sentences.

Linear Equations

An **equation** is a statement that says two mathematical expressions are equal. A **linear equation** in one variable is an equation with the exponent 1 on the variable. These are also know as **first degree equations,** because the highest exponent on the variable is 1. All linear equations can eventually be written in the form $ax + b = c$, where a, b, and c are real numbers and $a \neq 0$. It is assumed here that you are familiar with the addition and multiplication properties of equations.

■ *Addition property of equations:* If a, b, and c are real numbers and $a = b$, then $a + c = b + c$.

■ *Multiplication property of equations:* If a, b, and c are real numbers and $a = b$, then $ac = bc$.

The goal in solving linear equations is to get the variable isolated on either side of the equation by using the addition property of equations and then to get the coefficient of the variable to become 1 by using the multiplication property of equations.

Example 1: Solve for x: $6(2x - 5) = 4(8x + 7)$

$$6(2x - 5) = 4(8x + 7)$$

$$12x - 30 = 32x + 28$$

To isolate the x's on either side of the equation, you can either add $-12x$ to both sides or add $-32x$ to both sides.

$$
\begin{array}{r}
12x - 30 = 32x + 28 \\
\underline{-12x \qquad\quad -12x} \\
-30 = 20x + 28
\end{array}
$$

Isolate the $20x$.

$$
\begin{array}{r}
\underline{-28 \qquad\quad -28} \\
-58 = 20x
\end{array}
$$

Multiply each side by $\frac{1}{20}$.

$$\tfrac{1}{20}(-58) = \tfrac{1}{20}(20x)$$

$$-\tfrac{29}{10} = x$$

The solution is $-\frac{29}{10}$. This is indicated by putting the solution inside braces to form a set $\{-\frac{29}{10}\}$. This set is called the **solution set** of the equation. You can check this solution by replacing x with $-\frac{29}{10}$ in the original equation. The solution set is $\{-\frac{29}{10}\}$.

Example 2: Solve for x: $\dfrac{2x - 5}{4} - \dfrac{x}{3} = 2 - \dfrac{x + 4}{6}$

This equation will be made simpler to solve by first clearing fraction values. To do this, find the least common denominator (LCD) for all the denominators in the equation and multiply both sides of the equation by this value, using the distributive property.

$$\frac{2x-5}{4} - \frac{x}{3} = 2 - \frac{x+4}{6}$$

$$\text{LCD} = 12$$

$$12\left(\frac{2x-5}{4} - \frac{x}{3}\right) = 12\left(2 - \frac{x+4}{6}\right)$$

$$\overset{3}{\cancel{12}}\left(\frac{2x-5}{\cancel{4}}\right) - \overset{4}{\cancel{12}}\left(\frac{x}{\cancel{3}}\right) = 12\,(2) - \overset{2}{\cancel{12}}\left(\frac{x+4}{\cancel{6}}\right)$$

$$6x - 15 - 4x = 24 - 2x - 8$$

Don't forget that the -2 is distributed over *both* the x and the 4. Simplify both sides by combining like terms.

$$2x - 15 = 16 - 2x$$

Get the variable on one side. $\underline{+2x} \qquad\qquad \underline{+2x}$

$$4x - 15 = 16$$

Isolate the variable. $\underline{+15} \quad \underline{+15}$

$$4x = 31$$

Multiply by $\frac{1}{4}$. $\frac{1}{4}(4x) = \frac{1}{4}(31)$

$$x = \frac{31}{4}$$

You can check this for yourself. The solution set is $\{\frac{31}{4}\}$.

Formulas

A **formula** is an equation that describes a relationship between unknown values. Many problems are easily solved if the correct formula is known. To use formulas to solve problems, perform the following steps:

1. **Identify the appropriate formula.**
2. **Replace the variables in the formula with their known values.**
3. **Solve the formula for the remaining variable.**
4. **Check the solution in the original problem.**
5. **State the solution.**

Example 3: Find the length of a rectangle (see Figure 1-1) if its perimeter is 48 inches and its width is 8 inches.

Figure 1-1 Perimeter of a rectangle.

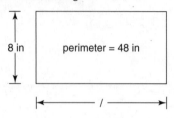

8 in   perimeter = 48 in

l

An appropriate formula is $p = 2l + 2w$, where p represents the perimeter, l the length, and w the width of the rectangle. Replace each variable with its known value.

$$p = 2l + 2w$$
$$48 = 2l + 2(8)$$

Now, solve for the remaining variable.

$$48 = 2l + 16$$

Add −16 to both sides.

$$32 = 2l$$

Multiply each side by $\frac{1}{2}$.

$$16 = l$$

Check with the original problem.

$$48 = 2l + 2(8)$$
$$48 \overset{?}{=} 2(16) + 2(8)$$
$$48 \overset{?}{=} 32 + 16$$
$$48 = 48 \checkmark$$

The length of the rectangle is 16 inches.

Example 4: As shown in Figure 1-2, an area rug is in the shape of a trapezoid. Its area is 38 square feet, its height is 4 feet, and one of its bases is 7 feet long. Find the length of the other base.

Figure 1-2 Area of trapezoid.

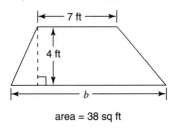

area = 38 sq ft

$$A_{\text{trapezoid}} = \tfrac{1}{2}\,h(b_1 + b_2)$$
$$38 = \tfrac{1}{2}(4)(7 + b_2)$$
$$38 = 2(7 + b_2)$$
$$38 = 14 + 2b_2$$
$$24 = 2b_2$$
$$12 = b_2$$

The check is left to you. The other base has a length of 12 feet.

Many times, a formula needs to be solved for one of its variables in terms of the other variables. To do this, simply perform the steps for solving equations as before but, this time, isolate the specified variable.

Example 5: $RT = D$ (Rate × Time = Distance). Solve for R.

$$RT = D$$
$$\tfrac{1}{T}\,(RT) = \tfrac{1}{T}\,(D)$$
$$R = \frac{D}{T}$$

Example 6: $A = P + PRT$ (Accumulated interest = Principal + Principal × Rate × Time). Solve for T.

$$A = P + PRT$$
$$A - P = PRT$$
$$\tfrac{1}{PR}\,(A - P) = \tfrac{1}{PR}\,(PRT)$$
$$\frac{A - P}{PR} = T$$

Absolute Value Equations

Recall that the absolute value of a number represents the distance that number is from zero on the number line. The equation $|x| = 3$ is translated as "x is 3 units from zero on the number line." Notice, on the number line shown in Figure 1-3, that there are two different numbers that are 3 units away from zero, namely, 3 or –3.

Figure 1-3   Absolute value.

The solution set of the equation $|x| = 3$ is {3, –3}, because $|3| = 3$ and $|-3| = 3$.

Example 7: Solve for x: $|4x - 2| = 8$

This translates to "$4x - 2$ is 8 units from zero on the number line" (see Figure 1-4).

Figure 1-4   There are + and – solutions.

$$\begin{array}{ccc}
4x - 2 = -8 & \text{or} & 4x - 2 = 8 \\
4x = -6 & & 4x = 10 \\
x = -\frac{3}{2} & & x = \frac{5}{2}
\end{array}$$

Check the solution.

$$|4x - 2| = 8 \qquad\qquad |4x - 2| = 8$$

$$\left|4\left(-\tfrac{3}{2}\right) - 2\right| \overset{?}{=} 8 \qquad\qquad \left|4\left(\tfrac{5}{2}\right) - 2\right| \overset{?}{=} 8$$

$$|-6 - 2| \overset{?}{=} 8 \qquad\qquad |10 - 2| \overset{?}{=} 8$$

$$|-8| \overset{?}{=} 8 \qquad\qquad |8| \overset{?}{=} 8$$

$$8 = 8 \checkmark \qquad\qquad 8 = 8 \checkmark$$

These are true statements. The solution set is $\{-\tfrac{3}{2}, \tfrac{5}{2}\}$.

Example 8: Solve for x: $\left|\tfrac{3x}{2} + 2\right| + 10 = 21$

To solve this type of absolute value equation, first isolate the expression involving the absolute value symbol.

$$\left|\tfrac{3x}{2} + 2\right| + 10 = 21$$

$$-10 \quad -10$$

$$\left|\tfrac{3x}{2} + 2\right| \qquad = 11$$

Now, translate the absolute value equation:

"$\tfrac{3x}{2} + 2$ is 11 units from zero on the number line."

$$\tfrac{3x}{2} + 2 = -11 \qquad \text{or} \qquad \tfrac{3x}{2} + 2 = 11$$

$$2\left(\tfrac{3x}{2} + 2\right) = 2(-11) \qquad\qquad 2\left(\tfrac{3x}{2} + 2\right) = 2(11)$$

$$3x + 4 = -22 \qquad\qquad 3x + 4 = 22$$

$$3x = -26 \qquad\qquad 3x = 18$$

$$x = -\tfrac{26}{3} \qquad\qquad x = 6$$

The check is left to you. The solution set is $\{-\tfrac{26}{3}, 6\}$.

Example 9: Solve for x: $|x| = -2$

This problem has no solutions, because the translation is nonsensical. Distance is not measured in negative values.

Example 10: Solve for x: $|2x - 3| = |3x + 7|$

This type of sentence will be true if either

1. The expressions *inside* the absolute value symbols are exactly the same (that is, they are equal) or

2. The expressions *inside* the absolute value symbols are opposites of each other.

$$2x - 3 = 3x + 7 \qquad \text{or} \qquad 2x - 3 = -(3x + 7)$$
$$-3 = x + 7 \qquad\qquad\qquad 2x - 3 = -3x - 7$$
$$-10 = x \qquad\qquad\qquad\qquad 5x - 3 = -7$$
$$5x = -4$$
$$x = -\tfrac{4}{5}$$

The check is left to you. The solution set is $\{-10, -\tfrac{4}{5}\}$.

Example 11: Solve for x: $|x - 2| = |7 - x|$

$$x - 2 = 7 - x \qquad \text{or} \qquad x - 2 = -(7 - x)$$
$$2x - 2 = 7 \qquad\qquad\qquad x - 2 = -7 + x$$
$$2x = 9 \qquad\qquad\qquad\qquad -2 = -7$$
$$x = \tfrac{9}{2}$$

The sentence $-2 = -7$ is never true, so it gives no solution.

Check the solution.

$$|x - 2| = |7 - x|$$

$$\left|\tfrac{9}{2} - 2\right| \overset{?}{=} \left|7 - \tfrac{9}{2}\right|$$

$$\left|\tfrac{9}{2} - \tfrac{4}{2}\right| \overset{?}{=} \left|\tfrac{14}{2} - \tfrac{9}{2}\right|$$

$$\left|\tfrac{5}{2}\right| \overset{?}{=} \left|\tfrac{5}{2}\right|$$

$$\tfrac{5}{2} = \tfrac{5}{2}$$

Therefore, the solution set is $\{\tfrac{9}{2}\}$.

Linear Inequalities

An **inequality** is a sentence using a symbol other than equality. The most common inequality symbols are $<$, $\le$, $>$, and $\ge$. To solve an inequality sentence, use exactly the same procedure as though it were an equation, with the following exception. When multiplying (or dividing) both sides of an inequality by a negative number, the direction of the inequality switches. This is called the **negative multiplication property of inequality.**

Negative Multiplication Property of Inequality

If a, b, and c are real numbers and c is negative, and $a < b$, then $ac > bc$. Or if $a > b$, then $ac < bc$.

Example 12: Solve for x: $3x - 7 > 20$

$$3x - 7 > 20$$

$$3x > 27$$

$$x > 9$$

To check the solution, first see whether $x = 9$ makes the equation $3x - 7 = 20$ true.

$$3x - 7 = 20$$

$$3(9) - 7 \overset{?}{=} 20$$

$$27 - 7 \overset{?}{=} 20$$

$$20 = 20 \checkmark$$

Now, choose a number greater than 9, 10 for example, and see if that makes the original inequality true.

$$3x - 7 > 20$$

$$3(10) - 7 \overset{?}{>} 20$$

$$30 - 7 \overset{?}{>} 20$$

$$23 > 20 \checkmark$$

This is a true statement. Since it is impossible to list all the numbers that are greater than 9, use "set builder" notation to show the solution set.

$$\{x \mid x > 9\}$$

This is read as "the set of all x so that x is greater than 9." Many times, the solutions to inequalities are graphed to illustrate the answers. The graph of $\{x \mid x > 9\}$ is shown in Figure 1-5.

Figure 1-5   Note that 9 is *not* included.

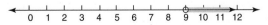

Example 13: Solve for x: $-\frac{5}{8}x + \frac{2}{3} > \frac{3}{4}x - \frac{7}{12}$

The LCD for this inequality is 24. Multiply both sides of the inequality by 24 as you would have had this been an equation.

$$24\left(-\tfrac{5}{8}x + \tfrac{2}{3}\right) > 24\left(\tfrac{3}{4}x - \tfrac{7}{12}\right)$$
$$-15x + 16 > 18x - 14$$

At this point, x can be isolated on either side of the inequality.

Isolating x on the left side,

$$-33x + 16 > -14$$
$$-33x > -30$$
$$x < \tfrac{10}{11}$$

Isolating x on the right side,

$$16 > 33x - 14$$
$$30 > 33x$$
$$\tfrac{10}{11} > x$$

In the final step on the left, the direction is switched because both sides are multiplied by a negative number. Both methods produce the final result that says that x is a number less than $\frac{10}{11}$. The check is left to you. The solution set is expressed as

$$\left\{x \,\middle|\, x < \tfrac{10}{11}\right\}$$

The graph of this solution set is shown in Figure 1-6.

Figure 1-6   Note the "hole" at $\frac{10}{11}$.

$$\frac{10}{11}$$

Compound Inequalities

A **compound inequality** is a sentence with two inequality statements joined either by the word "or" or by the word "and." "And" indicates that both statements of the compound sentence are true at the same time. It is the overlap or intersection of the solution sets for the individual statements. "Or" indicates that, as long as either statement is true, the entire compound sentence is true. It is the combination or union of the solution sets for the individual statements.

Example 14: Solve for x: $3x + 2 < 14$ and $2x - 5 > -11$

Solve each inequality separately. Since the joining word is "and," this indicates that the overlap or intersection is the desired result.

$$3x + 2 < 14 \qquad \text{and} \qquad 2x - 5 > -11$$
$$3x < 12 \qquad\qquad\qquad 2x > -6$$
$$x < 4 \qquad\qquad\qquad x > -3$$

$x < 4$ indicates all the numbers to the left of 4, and $x > -3$ indicates all the numbers to the right of -3. The intersection of these two graphs is all the numbers between -3 and 4. The solution set is

$$\left\{ x \mid x > -3 \text{ and } x < 4 \right\}$$

Another way this solution set could be expressed is

$$\left\{ x \mid -3 < x < 4 \right\}$$

When a compound inequality is written without the expressed word "and" or "or," it is understood to automatically be the word "and." Reading $\left\{ x \mid -3 < x < 4 \right\}$ from the "x" position, you say (reading to the left), "x is greater than -3 *and* (reading to the right) x is less than 4." The graph of the solution set is shown in Figure 1-7.

Figure 1-7 x is greater than -3 *and* less than 4.

Example 15: Solve for x: $2x + 7 < -11$ or $-3x - 2 < 13$

Solve each inequality separately. Since the joining word is "or," combine the answers; that is, find the union of the solution sets of each inequality sentence.

$$2x + 7 < -11 \qquad \text{or} \qquad -3x - 2 < 13$$
$$2x < -18 \qquad\qquad\qquad -3x < 15$$
$$x < -9 \qquad\qquad\qquad x > 5$$

Remember, as in the last step on the right, to switch the inequality when multiplying by a negative.

$x < -9$ indicates all the numbers to the left of -9, and $x > -5$ indicates all the numbers to the right of -5. The solution set is written as

$$\left\{ x \mid x < -9 \text{ or } x > -5 \right\}$$

The graph of this solution set is shown in Figure 1-8.

Figure 1-8 *x* is less than –9 and greater than –5.

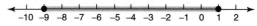

Example 16: Solve for *x*: $-12 \le 2x + 6 \le 8$

Since this compound inequality has no connecting word written, it is understood to be "and." It is translated into the following compound sentence.

$$-12 \le 2x + 6 \qquad \text{and} \qquad 2x + 6 \le 8$$
$$-18 \le 2x \qquad\qquad\qquad\qquad 2x \le 2$$
$$-9 \le x \qquad\qquad\qquad\qquad\quad x \le 1$$

$-9 \le x$ indicates all the numbers to the right of –9, and $x \le 1$ indicates all the numbers to the left of 1. The intersection of these graphs is the numbers between –9 and 1, including –9 and 1. The solution set can be written as

$$\left\{ x \mid x \ge -9 \text{ and } x \le 1 \right\} \quad \text{or} \quad \left\{ x \mid -9 \le x \le 1 \right\}$$

The graph of the solution set is shown in Figure 1-9.

Figure 1-9   Dots indicate inclusion of the points.

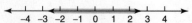

Example 17: Solve for *x*: $3x - 2 > -8$ or $2x + 1 < 9$

$$3x - 2 > -8 \qquad \text{or} \qquad 2x + 1 < 9$$
$$3x > -6 \qquad\qquad\qquad\qquad 2x < 8$$
$$x > -2 \qquad\qquad\qquad\qquad x < 4$$

$x > -2$ indicates all the numbers to the right of –2, and $x < 4$ indicates all the numbers to the left of 4. The union of these graphs is the entire number line. That is, the solution set is all real numbers. The graph of the solution set is the entire number line (see Figure 1-10).

Figure 1-10   Arrow heads indicate infinite.

Example 18: Solve for x: $4x - 2 < 10$ and $3x + 1 > 22$

$$4x - 2 < 10 \qquad \text{and} \qquad 3x + 1 > 22$$
$$4x < 12 \qquad\qquad\qquad 3x > 21$$
$$x < 3 \qquad\qquad\qquad\qquad x > 7$$

$x < 3$ indicates all the numbers to the left of 3, and $x > 7$ indicates all the numbers to the right of 7. The intersection of these graphs contains no numbers. That is, the solution set is the empty set, ϕ. A way to graph the empty set is to draw a number line but not to darken in any part of it. The graph of the empty set is shown in Figure 1-11.

Figure 1-11   The empty set.

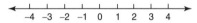

Absolute Value Inequalities

Remember, absolute value means distance from zero on a number line. $|x| < 4$ means that x is a number that is less than 4 units from zero on a number line (see Figure 1-12).

Figure 1-12   Less than 4 from 0.

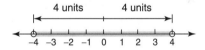

The solutions are the numbers to the right of -4 *and* to the left of 4 and could be indicated as

$$\left\{ x \mid x > -4 \text{ and } x < 4 \right\} \quad \text{or} \quad \left\{ x \mid -4 < x < 4 \right\}$$

$|x| > 4$ means that x is a number that is more than 4 units from zero on a number line (see Figure 1-13).

Figure 1-13 More than 4 from 0.

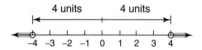

The solutions are the numbers to the left of -4 *or* to the right of 4 and are indicated as

$$\{x \mid x < -4 \text{ or } x > 4\}$$

$|x| < 0$ has no solutions, whereas $|x| > 0$ has as its solution all real numbers except 0. $|x| > -1$ has as its solution all real numbers, because after taking the absolute value of any number, that answer is either zero or positive and will always be greater than -1.

The following is a general approach for solving absolute value inequalities of the form

$$
\begin{array}{lll}
|ax + b| < c, & & |ax + b| > c, \\
|ax + b| \le c, & \text{or} & |ax + b| \ge c,
\end{array}
$$

■ If c is negative,

$|ax + b| < c$ has no solutions.

$|ax + b| \le c$ has no solutions.

$|ax + b| > c$ has as its solution all real numbers.

$|ax + b| \ge c$ has as its solution all real numbers.

■ If $c = 0$,

$|ax + b| < 0$ has no solutions.

$|ax + b| \le 0$ has as its solution the solution to $ax + b = 0$.

$|ax + b| > 0$ has as its solution all real numbers, except the solution to $ax + b = 0$.

$|ax + b| \ge 0$ has as its solution all real numbers.

■ If c is positive,

$|ax + b| < c$ has solutions that solve

$ax + b > -c$ and $ax + b < c$

$-c < ax + b < c$

That is,

$|ax + b| > c$ has solutions that solve

$$ax + b < -c \text{ or } ax + b > c$$

$|ax + b| \leq c$ has solutions that solve
$$-c \leq ax + b \leq c$$

$|ax + b| \geq c$ has solutions that solve
$$ax + b \leq -c \text{ or } ax + b \geq c$$

Example 19: Solve for x: $|3x - 5| < 12$

$$3x - 5 > 12 \text{ and } 3x - 5 < 12$$

$3x - 5 > -12$	and	$3x - 5 < 12$
$3x > -7$		$3x < 17$
$x > -\frac{7}{3}$		$x < \frac{17}{3}$

The solution set is

$$\left\{ x \,\middle|\, x > -\tfrac{7}{3} \text{ and } < \tfrac{17}{3} \right\} \text{ or } \left\{ x \,\middle|\, -\tfrac{7}{3} < x < \tfrac{17}{3} \right\}$$

The graph of the solution set is shown in Figure 1-14.

Figure 1-14 x is greater than $-\frac{7}{3}$ *and* less than $\frac{17}{3}$.

Example 20: Solve for x: $|5x + 3| > 2$

$$5x + 3 < -2 \text{ or } 5x + 3 > 2$$

$5x + 3 < -2$	or	$5x + 3 > 2$
$5x < -5$		$5x > -1$
$x < -1$		$x > -\frac{1}{5}$

The solution set is $\left\{ x \,\middle|\, x < -1 \text{ or } x > -\tfrac{1}{5} \right\}$. The graph of the solution set is shown in Figure 1-15.

Figure 1-15 x is less than -1 *or* greater than $-\frac{1}{5}$.

Example 21: Solve for x: $|2x + 11| < 0$

There is no solution for this inequality.

Example 22: Solve for x: $|2x + 11| > 0$.

The solution is all real numbers *except* for the solution to $2x + 11 = 0$. Therefore,

$$2x + 11 \neq 0$$

$$2x \neq -11$$

$$x \neq -\tfrac{11}{2}$$

The solution of the set is $\{x \mid x \text{ is a real number}, x \neq -\tfrac{11}{2}\}$. The graph of the solution set is shown in Figure 1-16.

Figure 1-16   All numbers but $-\tfrac{11}{2}$.

Example 23: Solve for x: $7|3x + 2| + 5 > 4$

First, isolate the expression involving the absolute value symbol.

$$7|3x + 2| + 5 > 4$$

$$7|3x + 2| > -1$$

$$|3x + 2| > -\tfrac{1}{7}$$

The solution set is all real numbers. (***Note:*** The absolute value of any number is always zero or a positive value. Therefore, the absolute value of any number is *always* greater than a negative value.) The graph of the solution set is shown in Figure 1-17.

Figure 1-17   The set of all numbers.

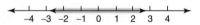

Chapter Checkout

Q&A

1. Solve for x:

$$5(3x - 7) = 6(5x + 9)$$

2. Final velocity (v_f) = initial velocity (v_i) + acceleration (a) × time (t). Solve the formula for a:

$$v_f = v_i + at$$

3. Solve for y:

$$3y + 2 < 17 \quad \text{and} \quad 2y - 3 > -15$$

4. Solve for x:

$$|5x - 3| < 14$$

Answers: 1. $x = -\frac{89}{15}$ 2. $a = \frac{v_f - v_i}{t}$ 3. $\left\{ y \middle| -6 < y < 5 \right\}$ 4. $\left\{ x \middle| -\frac{11}{5} < x < \frac{17}{5} \right\}$

Chapter 2

SEGMENTS, LINES, AND INEQUALITIES

Chapter Check-In

❑ Learning the rectangular coordinate system

❑ Deriving and applying the distance formula

❑ Understanding and applying the midpoint formula

❑ Understanding slope and intercepts

❑ Graphing linear equations and inequalities

Equations can be graphed on a set of coordinate axes. The location of every point on a graph can be determined by two coordinates, written as an ordered pair, (x,y). These are also known as Cartesian coordinates, after the French mathematician Rene Descartes, who is credited with their invention. If the slope and intercept or the coordinates of two points on a linear graph are known, the equation of the line can be determined.

Rectangular Coordinate System

Following are terms you should be familiar with:

■ **Coordinates of a point.** Each point on a number line is assigned a number. In the same way, each point in a plane is assigned a pair of numbers called the **coordinates of the point.**

■ **x-axis; y-axis.** To locate points in a plane, two perpendicular lines are used: a horizontal line called the **x-axis** and a vertical line called the **y-axis.**

■ **Origin.** The point of intersection of the x-axis and the y-axis is called the **origin.**

■ **Coordinate plane.** The x-axis, the y-axis, and all the points in their plane are called a **coordinate plane.**

■ **Ordered pairs.** Every point in a coordinate plane is named by a pair of numbers whose order is important. This pair of numbers, written in parentheses and separated by a comma, is the **ordered pair** for the point. The ordered pair for the origin is (0,0).

■ **x-coordinate.** The number to the left of the comma in an ordered pair is the **x-coordinate** of the point and indicates the amount of movement along the x-axis from the origin. The movement is to the right if the number is positive and to the left if the number is negative.

■ **y-coordinate.** The number to the right of the comma in an ordered pair is the **y-coordinate** of the point and indicates the amount of movement perpendicular to the x-axis. The movement is above the x-axis if the number is positive and below the x-axis if the number is negative.

■ **Quadrants.** The x-axis and y-axis separate the coordinate plane into four regions called **quadrants.** The upper-right quadrant is quadrant I, the upper-left quadrant is quadrant II, the lower-left quadrant is quadrant III, and the lower-right quadrant is quadrant IV. Notice that, as shown in Figure 2-1,

in quadrant **I,** x is always positive and y is always positive (+,+)

in quadrant **II,** x is always negative and y is always positive (−,+)

in quadrant **III,** x is always negative and y is always negative (−,−)

in quadrant **IV,** x is always positive and y is always negative (+,−)

Figure 2-1 The quadrants.

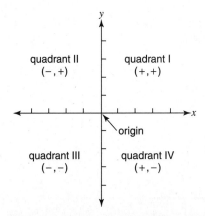

The point associated with an ordered pair of real numbers is called the **graph** of the ordered pair.

Distance Formula

In Figure 2-2, *A* is (2,2), *B* is (5,2), and *C* is (5,6).

Figure 2-2   The distance formula.

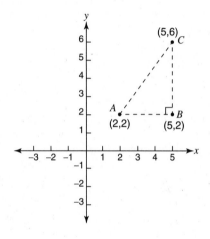

To find the length of *AB* or *BC*, only simple subtraction is necessary.

$$AB = 5 - 2 \qquad \text{and} \qquad BC = 6 - 2$$
$$= 3 \qquad\qquad\qquad\qquad = 4$$

To find the length of *AC*, however, simple subtraction is not sufficient. Triangle *ABC* is a right triangle with *AC* the hypotenuse. Therefore, by the Pythagorean theorem,

$$AC^2 = AB^2 + BC^2$$
$$AC = \sqrt{AB^2 + BC^2}$$
$$= \sqrt{3^2 + 4^2}$$
$$= \sqrt{9 + 16}$$
$$= \sqrt{25}$$
$$= 5$$

If A is represented by the ordered pair (x_1, y_1) and C is represented by the ordered pair (x_2, y_2), then

$$AB = |x_2 - x_1| \quad \text{and} \quad BC = |y_2 - y_1|$$

Then

$$AC = \sqrt{(x_2 - x_1)^2 + (y_2 - y_1)^2}$$

> *Distance Formula*
>
> $$d = \sqrt{(x_2 - x_1)^2 + (y_2 - y_1)^2}$$

Example 1: Use the distance formula to find the distance between the points with coordinates $(-3, 4)$ and $(5, 2)$.

Let $(-3, 4) = (x_1, y_1)$ and $(5, 2) = (x_2, y_2)$. Then

$$
\begin{aligned}
d &= \sqrt{[5 - (-3)]^2 + (2 - 4)^2} \\
&= \sqrt{8^2 + (-2)^2} \\
&= \sqrt{64 + 4} \\
&= \sqrt{68} \\
&= \sqrt{4}\sqrt{17} \\
&= 2\sqrt{17}
\end{aligned}
$$

Midpoint Formula

Numerically, the midpoint of a segment can be considered to be the average of its endpoints. This concept helps in remembering a formula for finding the midpoint of a segment, given the coordinates of its endpoints. Recall that the average of two numbers is found by dividing their sum by two.

> *Midpoint Formula*
>
> $$M = \left(\frac{x_1 + x_2}{2}, \frac{y_1 + y_2}{2} \right)$$

Example 2: In Figure 2-3, R is the midpoint between $Q(-9, -1)$ and $T(-3, 7)$. Find its coordinates.

Figure 2-3   Finding the midpoint.

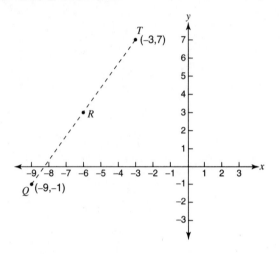

By the midpoint formula,

$$M = \left(\frac{x_1 + x_2}{2}, \frac{y_1 + y_2}{2} \right)$$

$$R = \left(\frac{-9 + (-3)}{2}, \frac{-1 + 7}{2} \right)$$

$$= \left(\frac{-12}{2}, \frac{6}{2} \right)$$

$$= (-6, 3)$$

Slope of a Line

The **slope of a line** is a measurement of the steepness and direction of a nonvertical line. When a line slants from lower left to upper right, the slope is a positive number. Item (a) in Figure 2-4 shows a line with a positive slope. When a line slants from lower right to upper left, the slope is a negative number (b). The x- axis or any line parallel to the x-axis has a slope of zero; that is, a horizontal line has a slope of zero (c). The y-axis or any line parallel to the y-axis has no defined slope; that is, a vertical line has an undefined slope (d).

Figure 2-4 Slopes of lines.

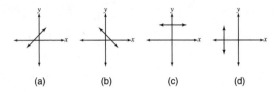

(a)   (b)   (c)   (d)

If m represents the slope of a line and A and B are points with coordinates (x_1, y_1) and (x_2, y_2), respectively, then the slope of the line passing through A and B is given by the following formula.

> *Slope Formula*
> $$m = \frac{y_2 - y_1}{x_2 - x_1}, \qquad x_2 \neq x_1$$

Since A and B cannot be points on a vertical line, x_1 and x_2 cannot be equal to one another. If $x_1 = x_2$, then the line is vertical, and the slope is undefined.

Example 3: Use Figure 2-5 to find the slopes of the lines a, b, c, and d.

Figure 2-5 Find the slopes.

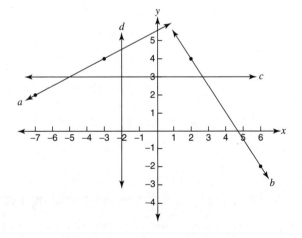

Line a passes through the points $(-7,2)$ and $(-3,4)$.

$$m = \frac{y_2 - y_1}{x_2 - x_1}$$

$$= \frac{4 - 2}{-3 - (-7)}$$

$$= \frac{2}{4}$$

$$= \frac{1}{2}$$

Line *b* passes through the points (2,4) and (6,–2).

$$m = \frac{y_2 - y_1}{x_2 - x_1}$$

$$= \frac{-2 - 4}{6 - 2}$$

$$= \frac{-6}{4}$$

$$= -\frac{3}{2}$$

Line *c* is parallel to the *x*-axis. Therefore,

$$m = 0$$

Line *d* is parallel to the *y*-axis. Therefore, line *d* has an undefined slope.

Example 4: A line passes through (–5,8) with a slope of $\frac{2}{3}$. If another point on this line has coordinates (*x*,12), find *x*.

$$m = \frac{y_2 - y_1}{x_2 - x_1}$$

$$\frac{2}{3} = \frac{12 - 8}{x - (-5)}$$

$$\frac{2}{3} = \frac{4}{x + 5}$$

$$2(x + 5) = 4(3)$$

$$2x + 10 = 12$$

$$2x = 2$$

$$x = 1$$

Slope of Parallel and Perpendicular Lines

Parallel lines have equal slopes. Conversely, if two different lines have equal slopes, they are parallel. If two nonvertical lines are perpendicular, then their slopes are negative reciprocals (actually, opposite reciprocals) of one another, or the product of their slopes is –1. Conversely, if the slopes of two lines are opposite reciprocals of one another, or the product of their slopes is –1, then the lines are nonvertical perpendicular lines. Because horizontal and vertical lines are always perpendicular, then lines having a zero slope and an undefined slope will be perpendicular.

Example 5: If line l has slope $\frac{3}{4}$, then

       a)   any line parallel to line l will have slope _____, and

       b)   any line perpendicular to line l will have slope _____.

           (a)$\frac{3}{4}$ (b)$-\frac{4}{3}$

Equations of Lines

Equations involving one or two variables can be graphed on any x-y coordinate plane. In general, it is true that

■ if a point lies on the graph of an equation, then its coordinates make the equation a true statement, and

■ if the coordinates of a point make an equation a true statement, then the point lies on the graph of the equation.

The graphs of linear equations are always lines. All linear equations can be written in the form $Ax + By = C$, where A, B, and C are real numbers and A and B are not both zero. Futhermore, to be in standard form, A has to be a positive number. Below are examples of linear equations and their respective A, B, and C values.

$x + y = 0$	$3x - 4y = 9$	$x = -6$	$y = 7$
$A = 1$	$A = 3$	$A = 1$	$A = 0$
$B = 1$	$B = -4$	$B = 0$	$B = 1$
$C = 0$	$C = 9$	$C = -6$	$C = 7$

Following are terms you should be familiar with:

■ **Standard form.** The form $Ax + By = C$ for the equation of a line is known as the **standard form** for the equation of a line.

■ **x-intercept.** The **x-intercept** of a graph is the point at which the graph will intersect the x-axis. It will always have a y-coordinate of zero. A horizontal line that is not the x-axis will have no x-intercept.

■ **y-intercept.** The **y-intercept** of a graph is the point at which the graph will intersect the *y*-axis. It will always have an *x*-coordinate of zero. A vertical line that is not the *y*-axis will have no *y*-intercept.

One way to graph a linear equation is to find solutions by giving a value to one variable and solving the resulting equation for the other variable. A minimum of two points is necessary to graph a linear equation.

Example 6: Draw the graph of $2x + 3y = 12$ by finding two random points.

To do this, select a value for one variable; then substitute this into the equation and solve for the other variable. Do this a second time with new values to get a second point.

Let $x = 2$; then find *y*.

$$2x + 3y = 12$$
$$2(2) + 3y = 12$$
$$4 + 3y = 12$$
$$3y = 8$$
$$y = \frac{8}{3}$$

Therefore, the ordered pair $(2, \frac{8}{3})$ belongs on the graph.

Let $y = 6$; then find *x*.

$$2x + 3y = 12$$
$$2x + 3(6) = 12$$
$$2x + 18 = 12$$
$$2x = -6$$
$$x = -3$$

Therefore, the ordered pair $(-3,6)$ belongs on the graph.

As shown in Figure 2-6, graph these points and then connect them to make the line that represents the graph of $2x + 3y = 12$.

Figure 2-6 $2x + 3y = 12.$

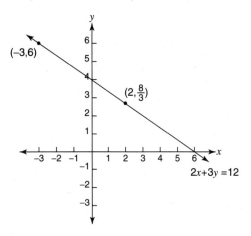

Example 7: Draw the graph of $2x + 3y = 12$ by finding the x-intercept and the y-intercept.

The x-intercept has a y-coordinate of zero. Substituting zero for y, the resulting equation is

$$2x + 3(0) = 12$$

Now, solving for x,

$$2x = 12$$
$$x = 6$$

The x-intercept is at $(6,0)$.

The y-intercept has an x-coordinate of zero. Substituting zero for x, the resulting equation is

$$2(0) + 3y = 12$$

Now, solving for y,

$$3y = 12$$
$$y = 4$$

The y-intercept is at $(0,4)$.

The line can now be graphed, as shown in Figure 2-7, by plotting these two points and drawing the line they determine.

Figure 2-7 *x*- and *y*- intercepts.

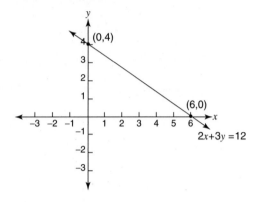

Notice that Figures 2-6 and 2-7 are exactly the same. Both are the graph of the line $2x + 3y = 12$.

Example 8: Draw the graph of $x = 2$.

As shown in Figure 2-8, $x = 2$ is a vertical line whose *x*-coordinate is always 2.

Figure 2-8 $x = 2$ for all *y* values.

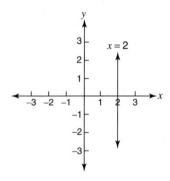

Example 9: Draw the graph of $y = -1$.

As shown in Figure 2-9, $y = -1$ is a horizontal line whose *y*-coordinate is always -1.

Figure 2-9 $y = -1$ for all x values.

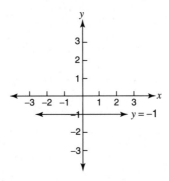

Suppose that A is a particular point called (x_1, y_1) and B is any point called (x, y). Then the slope of the line through A and B is represented by

$$\frac{y - y_1}{x - x_1} = m$$

Apply the cross products property, and the equation becomes

$$y - y_1 = m(x - x_1)$$

This is the **point-slope form** of a nonvertical line.

Example 10: Find the equation of the line containing the points $(-3, 4)$ and $(7, 2)$ and write the equation in both point-slope form and standard form.

For the point-slope form, first find the slope, m.

$$\frac{y - y_1}{x - x_1} = m$$

$$m = \frac{2 - 4}{7 - (-3)}$$

$$= -\frac{2}{10}$$

$$= -\frac{1}{5}$$

Now, choose either given point, say $(-3, 4)$, and substitute the x and y values into the point-slope form.

$$y - y_1 = m(x - x_1)$$

$$y - 4 = -\tfrac{1}{5}\,[(x - (-3)] \text{ or } y - 4 = -\tfrac{1}{5}\,(x + 3)$$

For the standard form, begin with the point-slope form and clear it of fractions by multiplying both sides by the least common denominator.

$$y - 4 = -\tfrac{1}{5}(x + 3)$$

Multiply both sides by 5.

$$5(y - 4) = 5[-\tfrac{1}{5}(x + 3)]$$
$$5y - 20 = -(x + 3)$$
$$5y - 20 = -x - 3$$

Get x and y on one side and the constants on the other side by adding x to both sides and adding 20 to both sides. Make sure A is a positive number.

$$x + 5y = 17$$

A nonvertical line written in standard form is $Ax + By = C$, with $B \neq 0$. Solve this equation for y.

$$Ax + By = C$$
$$By = -Ax + C$$
$$y = -\tfrac{A}{B}x + \tfrac{C}{B}$$

The value $-\tfrac{A}{B}$ becomes the slope of the line, and $\tfrac{C}{B}$ becomes the y-intercept value. If $-\tfrac{A}{B}$ is replaced with m and $\tfrac{C}{B}$ is replaced with b, the equation becomes $y = mx + b$. This is known as the **slope-intercept form** of a nonvertical line.

Example 11: Find the slope and y-intercept value of the line with the equation $3x - 4y = 20$.

Solve for y.

$$3x - 4y = 20$$
$$-4y = -3x + 20$$
$$y = \tfrac{3}{4}x - 5$$

Therefore, the slope of the line is $\tfrac{3}{4}$, and the y-intercept value is -5.

Example 12: Draw the graph of the equation $y = -\tfrac{3}{4}x + 5$.

The equation is in slope-intercept form. The slope is $-\tfrac{3}{4}$, and the y-intercept is at $(0,5)$. From this, the graph can be quickly drawn. Because the slope is negative, the line is slanting to the upper left/lower right. Begin with the y-intercept $(0,5)$ and use the slope to find additional points. Either go up 3 and left 4 or go down 3 and right 4. Now, label these points and connect them. Look at Figure 2-10.

Figure 2-10 Slope-intercept form.

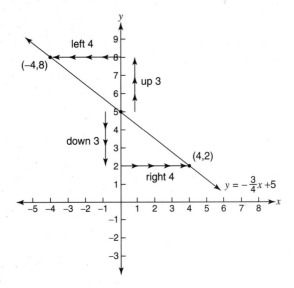

Graphs of Linear Inequalities

A **linear inequality** is a sentence in the form

$$Ax + By < C, Ax + By > C, Ax + By \leq C, \text{ or } Ax + By > C$$

To graph such sentences

1. **Graph the linear equation $Ax + By = C$.** This line becomes a boundary line for the graph. If the original inequality is < or >, the boundary line is drawn as a dashed line, since the points on the line do *not* make the original sentence true. If the original inequality is ≤ or ≥, the boundary line is drawn as a solid line, since the points on the line will make the original inequality true.

2. **Select a point not on the boundary line and substitute its x and y values into the original inequality.**

3. **Shade the appropriate area.** If the resulting sentence is true, then shade the region where that test point is located, indicating that all the points on that side of the boundary line will make the original sentence true. If the resulting sentence is false, then shade the region on the side of the boundary line opposite to where the test point is located.

Example 13: Graph $3x + 4y < 12$.

First, draw the graph of $3x + 4y = 12$. If you use the x-intercept and y-intercept method, you get x-intercept $(4, 0)$ and y-intercept $(0, 3)$. If you use the slope-intercept method, the equation, when written in $y = mx + b$ form becomes

$$y = -\tfrac{3}{4}x + 3$$

Because the original inequality is <, the boundary line will be a dashed line. Look at Figure 2-11.

Figure 2-11 The boundary is dashed.

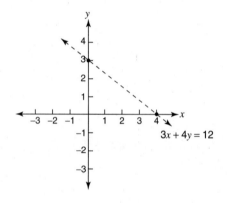

Now, select a point not on the boundary, say $(0, 0)$. Substitute this into the original inequality:

$$3x + 4y < 12$$
$$3(0) + 4(0) \overset{?}{<} 12$$
$$0 + 0 \overset{?}{<} 12$$
$$0 < 12 \ \checkmark$$

This is a true statement. This means that the "(0, 0) side" of the boundary line is the desired region to be shaded. Now, shade that region as shown in Figure 2-12.

Figure 2-12 The shading is below the line.

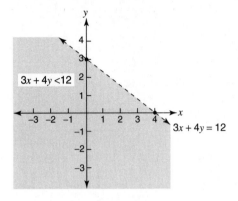

Example 14: Graph $y \geq 2x + 3$.

First, graph $y = 2x + 3$ (see Figure 2-13).

Figure 2-13 This boundary is solid.

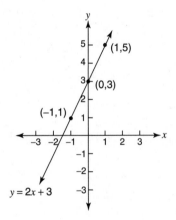

Notice that the boundary is a solid line, because the original inequality is $\geq$. Now, select a point not on the boundary, say (2, 1), and substitute its x and y values into $y \geq 2x + 3$.

$$y \geq 2x + 3$$
$$1 \overset{?}{\geq} 2(2) + 3$$
$$1 \overset{?}{\geq} 4 + 3$$
$$1 \geq 7 \qquad \text{No}$$

This is not a true statement. Because this replacement does not make the original sentence true, shade the region on the opposite side of the boundary line (see Figure 2-14).

Figure 2-14   Shading shows greater than.

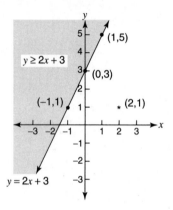

Example 15: Graph $x < 2$.

The graph of $x = 2$ is a vertical line whose points all have the x-coordinate of 2 (see Figure 2-15).

Figure 2-15 Dotted graph of $x = 2$.

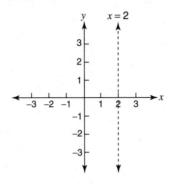

Select a point not on the boundary, say (0, 0). Substitute the x value into $x < 2$.

$$x < 2$$
$$0 < 2 \checkmark$$

This is a true statement. Therefore, shade in the "(0, 0) side" of the boundary line (see Figure 2-16).

Figure 2-16 x less than 2 is shaded.

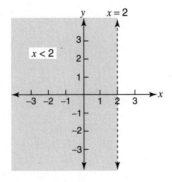

Chapter Checkout

Q&A

1. A line segment connects (1, 3) and (6, 7). Using the distance formula, find the length of that segment.

2. A line segment connects A (-5, -9) with B (7, 21). Point R is at the midpoint of that line. What are R's coordinates?

3. A line passes through (-5, 12) and (4, 7). What is its slope?

4. What are the coordinates of the x- and y-intercepts in the following equation?

$$3x + 2y = 24$$

Answers: 1. $\sqrt{41}$ 2. $(1, 6)$ 3. $-\frac{5}{9}$ 4. $(8, 0), (0, 12)$

Chapter 3

LINEAR SENTENCES IN TWO VARIABLES

Chapter Check-In

❑ Graphing to solve equations and inequalities

❑ Solving systems of equations by substitution

❑ Solving systems of equations by elimination

❑ Using matrices to solve systems of equations

❑ Using determinants to solve equation systems

Earlier chapters cover solving equations with a single variable. This chapter looks at ways to solve equations that contain two variables; these ways include graphing, substitution, elimination, matrices, and determinants. Certain sets of equations are more easily solved using one method instead of another.

Linear Equations: Solutions Using Graphing

Example 1: Solve this system of equations by using graphing.

$$\begin{cases} 4x + 3y = 6 \\ 2x - 5y = 16 \end{cases}$$

To solve using graphing, graph both equations on the same set of coordinate axes and see where the graphs cross. The ordered pair at the point of intersection becomes the solution (see Figure 3-1).

Figure 3-1 Two linear equations.

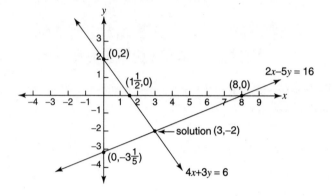

Check the solution.

$$4x + 3y = 6 \qquad\qquad 2x - 5y = 16$$

$$4(3) + 3(-2) \stackrel{?}{=} 6 \qquad 2(3) - 5(-2) \stackrel{?}{=} 16$$

$$12 - 6 \stackrel{?}{=} 6 \qquad\qquad 6 + 10 \stackrel{?}{=} 16$$

$$6 = 6 \checkmark \qquad\qquad 16 = 16 \checkmark$$

The solution is $x = 3$, $y = -2$.

Here are two things to keep in mind:

■ **Dependent system.** If the two graphs coincide—that is, if they are actually two versions of the same equation—then the system is called a **dependent system,** and its solution can be expressed as either of the two original equations.

■ **Inconsistent system.** If the two graphs are parallel—that is, if there is no point of intersection—then the system is called an **inconsistent system,** and its solution is expressed as an empty set, or null set.

Linear Equations: Solutions Using Substitution

To solve systems using substitution, follow this procedure.

1. **Select one equation and solve for one of its variables.**
2. **In the other equation, substitute for the variable just solved.**
3. **Solve the new equation.**
4. **Substitute the value found into any equation involving both variables and solve for the other variable.**
5. **Check the solution in both original equations.**

Example 2: Solve this system of equations by using substitution.

$$\begin{cases} 3x + 4y = -5 \\ 2x - 3y = 8 \end{cases}$$

Solve for x.

$$2x - 3y = 8$$
$$2x = 3y + 8$$
$$x = \tfrac{3}{2} y + 4$$

Substitute $\tfrac{3}{2} y + 4$ for x in the other equation.

$$3x + 4y = -5$$
$$3\left(\tfrac{3}{2} y + 4\right) + 4y = -5$$

Solve this new equation.

$$\tfrac{9}{2} y + 12 + 4y = -5$$
$$\tfrac{17}{2} y = -17$$
$$y = -2$$

Substitute into any equation involving both variables.

$$x = \frac{3}{2}y + 4$$

$$= \frac{3}{2}(-2) + 4$$

$$= -3 + 4$$

$$= 1$$

Check the solution in both original equations.

$$3x + 4y = -5 \qquad\qquad 2x - 3y = 8$$

$$3(1) + 4(-2) \overset{?}{=} -5 \qquad\qquad 2(1) - 3(-2) \overset{?}{=} 8$$

$$3 - 8 \overset{?}{=} -5 \qquad\qquad 2 + 6 \overset{?}{=} 8$$

$$-5 = -5 \checkmark \qquad\qquad 8 = 8 \checkmark$$

The solution is $x = 1$, $y = -2$.

If the substitution method produces a sentence that is always true, such as $0 = 0$, then the system is dependent, and either original equation is a solution. If the substitution method produces a sentence that is always false, such as $0 = 5$, then the system is inconsistent, and there is no solution.

Linear Equations: Solutions Using Eliminations

To solve systems using elimination, follow this procedure.

1. **Arrange both equations in standard form with like terms above one another.**
2. **Choose a variable to eliminate, and with a proper choice of multiplication, arrange so that the coefficients of that variable are opposites of one another.**
3. **Add the equations, leaving one equation with one variable.**
4. **Solve for the remaining variable.**

5. **Either repeat Steps 2 through 4, eliminating the other variable, or substitute the value found in Step 4 into any equation involving both variables and solve for the other variable.**

6. **Check the solution in both original equations.**

Example 3: Solve this system of equations by using elimination.

$$\begin{cases} 2x - 3 = -5y \\ -2y = -3x + 1 \end{cases}$$

Arrange both equations in standard form, putting like terms above one another.

$$2x + 5y = 3$$

$$3x - 2y = 1$$

Select a variable to eliminate, say y.

The coefficients of y are 5 and –2. These both divide into 10. Arrange so that the coefficient of y is 10 in one equation and –10 in the other. To do this, multiply the top equation by 2 and the bottom equation by 5.

$$2x + 5y = 3 \xrightarrow{\text{multiply}(2)} 4x + 10y = 6$$

$$3x - 2y = 1 \xrightarrow{\text{multiply}(5)} 15x - 10y = 5$$

Add the new equations, eliminating y.

$$4x + 10y = 6$$
$$\underline{15x - 10y = 5}$$
$$19x \qquad = 11$$

Solve for the remaining variable.

$$x = \tfrac{11}{19}$$

Repeat Steps 2 through 4 (using new multiplication factors) to eliminate x and solve for y or substitute the above value for x and solve for y.

Method 1: Eliminate x and solve for y.

$$6x + 15y = 9$$

$$2x + 5y = 3 \xrightarrow{\text{multiply}(3)}$$

$$3x - 2y = 1 \xrightarrow{\text{multiply}(-2)}$$

$$\underline{-6x + 4y = -2}$$

$$19y = 7$$

$$y = \tfrac{7}{19}$$

Method 2: Substitute for x and solve for y.

$$2x + 5y = 3$$

$$2\left(\tfrac{11}{19}\right) + 5y = 3$$

$$\tfrac{22}{19} + 5y = \tfrac{57}{19}$$

$$5y = \tfrac{35}{19}$$

$$y = \tfrac{7}{19}$$

Check the solution in the original equation.

$$2x - 3 = -5y \qquad\qquad\qquad -2y = -3x + 1$$

$$2\left(\tfrac{11}{19}\right) - 3 \overset{?}{=} -5\left(\tfrac{7}{19}\right) \qquad -2\left(\tfrac{7}{19}\right) \overset{?}{=} -3\left(\tfrac{11}{19}\right) + 1$$

$$\tfrac{22}{19} - \tfrac{57}{19} \overset{?}{=} -\tfrac{35}{19} \qquad\qquad -\tfrac{14}{19} \overset{?}{=} -\tfrac{33}{19} + \tfrac{19}{19}$$

$$-\tfrac{35}{19} = -\tfrac{35}{19}\ \checkmark \qquad\qquad\qquad -\tfrac{14}{19} = -\tfrac{14}{19}\ \checkmark$$

These are both true statements. The solution is $x = \tfrac{11}{19}$, $y = \tfrac{7}{19}$.

If the elimination method produces a sentence that is always true, then the system is dependent, and either original equation is a solution. If the elimination method produces a sentence that is always false, then the system is inconsistent and there is no solution.

Linear Equations: Solutions Using Matrices

A **matrix** (plural, **matrices**) is a rectangular array of numbers or variables. A matrix can be used to represent a system of equations in standard form by writing only the coefficients of the variables and the constants in the equations.

Example 4: Represent this system as a matrix.

$$\begin{cases} 5x - 2y = 13 \\ 2x + y = 7 \end{cases}$$

$$\left.\begin{array}{r} 5x - 2y = 13 \\ 2x + y = 7 \end{array}\right\} \longrightarrow \begin{bmatrix} 5 & -2 & \vdots & 13 \\ 2 & 1 & \vdots & 7 \end{bmatrix}$$

In the matrix above, the dashed line separates the coefficients of the variables from the constants in each equation.

Through the use of row multiplication and row additions, the goal is to transform the preceding matrix into the following form.

$$\begin{bmatrix} a & b & \vdots & c \\ 0 & d & \vdots & e \end{bmatrix}$$

The matrix method is the same as the elimination method but more organized.

Example 5: Solve this system by using matrices.

$$\begin{cases} 5x - 2y = 13 \\ 2x + y = 7 \end{cases}$$

$$\left.\begin{array}{r} 5x - 2y = 13 \\ 2x + y = 7 \end{array}\right\} \rightarrow \begin{bmatrix} 5 & -2 & \vdots & 13 \\ 2 & 1 & \vdots & 7 \end{bmatrix} \begin{array}{l} \text{multiply}(2) \\ \text{multiply}(-5) \end{array}$$

$$\begin{array}{r} \text{Rewrite row } 1\rightarrow \\ \text{Write the sum of 2 times row 1 and } -5 \text{ times row } 2\rightarrow \end{array} \begin{bmatrix} 5 & -2 & \vdots & 13 \\ 0 & -9 & \vdots & -9 \end{bmatrix}$$

This matrix now represents the system

$$5x - 2y = 13$$
$$-9y = -9$$

Therefore, $y = 1$

Now, substitute 1 for y in the other equation and solve for x.

$$5x - 2y = 13$$
$$5x - 2(1) = 13$$
$$5x = 15$$
$$x = 3$$

Check the solution.

$$5x - 2y = 13 \qquad\qquad 2x + y = 7$$
$$5(3) - 2(1) \overset{?}{=} 13 \qquad\qquad 2(3) + 1 \overset{?}{=} 7$$
$$15 - 2 \overset{?}{=} 13 \qquad\qquad 6 + 1 \overset{?}{=} 7$$
$$13 = 13 \checkmark \qquad\qquad 7 = 7 \checkmark$$

The solution is $x = 3$, $y = 1$.

Matrices will be more useful when solving a system of three equations in three unknowns.

Linear Equations: Solutions Using Determinants

A square array of numbers or variables enclosed between vertical lines is called a **determinant.** A determinant is different from a matrix in that a determinant has a numerical value, whereas a matrix does not. The following determinant has two rows and two columns.

$$\text{row 1} \longrightarrow \begin{vmatrix} a & c \\ b & d \end{vmatrix} \longleftarrow \text{row 2}$$

$$\underset{\substack{\text{column} \\ 1}}{\uparrow} \quad \underset{\substack{\text{column} \\ 2}}{\uparrow}$$

The value of this determinant is found by finding the difference between the diagonally down product and the diagonally up product:

$$\begin{vmatrix} a & c \\ b & d \end{vmatrix} = ad - bc$$

Example 6: Evaluate the following determinant.

$$\begin{vmatrix} 3 & -11 \\ 7 & 2 \end{vmatrix}$$

$$\begin{vmatrix} 3 & -11 \\ 7 & 2 \end{vmatrix} = (3)(2) - (7)(-11)$$

$$= 6 - (-77)$$

$$= 6 + 77$$

$$= 83$$

Example 7: Solve the following system by using determinants.

$$\begin{cases} 4x - 3y = -14 \\ 3x - 5y = -5 \end{cases}$$

To solve this system, three determinants will be created. One is called the **denominator determinant,** labeled D; another is the **x-numerator determinant,** labeled D_x; and the third is the **y-numerator determinant,** labeled D_y.

The denominator determinant, D, is formed by taking the coefficients of x and y from the equations written in standard form.

$$D = \begin{vmatrix} 4 & -3 \\ 3 & -5 \end{vmatrix} \qquad \begin{aligned} &= (4)(-5) - (3)(-3) \\ &= -20 - (-9) \\ &= -20 + 9 \\ &= -11 \end{aligned}$$

The *x*-numerator determinant is formed by taking the constant terms from the system and placing them in the *x*-coefficient positions and retaining the *y*-coefficients.

$$D_x = \begin{vmatrix} -14 & -3 \\ -5 & -5 \end{vmatrix} \qquad \begin{aligned} &= (-14)(-5) - (-5)(-3) \\ &= 70 - 15 \\ &= 55 \end{aligned}$$

The *y*-numerator determinant is formed by taking the constant terms from the system and placing them in the *y*-coefficient positions and retaining the *x*-coefficients.

$$D_y = \begin{vmatrix} 4 & -14 \\ 3 & -5 \end{vmatrix} \qquad \begin{aligned} &= (4)(-5) - (3)(-14) \\ &= -20 - (-42) \\ &= -20 + 42 \\ &= 22 \end{aligned}$$

The answers for *x* and *y* are

$$x = \frac{D_x}{D} = \frac{55}{-11} = -5 \qquad\qquad y = \frac{D_y}{D} = \frac{22}{-11} = -2$$

The check is left to you. The solution is $x = -5$, $y = -2$.

Many times, finding solutions by using determinants is referred to as **Cramer's rule,** named after the mathematician who devised this method.

Example 8: Use Cramer's rule to solve this system.

$$\begin{cases} 4x + 6y = 3 \\ 8x - 3y = 1 \end{cases}$$

$$D = \begin{vmatrix} 4 & 6 \\ 8 & -3 \end{vmatrix} = -12 - 48 = -60$$

$$D_x = \begin{vmatrix} 3 & 6 \\ 1 & -3 \end{vmatrix} = -9 - 6 = -15$$

$$D_y = \begin{vmatrix} 4 & 3 \\ 8 & 1 \end{vmatrix} = 4 - 24 = -20$$

$$x = \frac{D_x}{D} = \frac{-15}{-60} = \frac{1}{4}, \qquad y = \frac{D_y}{D} = \frac{-20}{-60} = \frac{1}{3}$$

The check is left to you. The solution is $x = \frac{1}{4}$, $y = \frac{1}{3}$.

Linear Inequalities: Solutions Using Graphing

Example 9: Graph the solution to this system of inequalities.

$$\begin{cases} x + 2y \le 12 \\ 3x - y \ge 6 \end{cases}$$

To graph the solution to a system of inequalities, follow this procedure:

1. **Graph each sentence on the same set of axes.**
2. **See where the shading of the sentences overlaps.**

   The overlapping region is the solution to the system of inequalities. The solution to the system is the region with both shadings (see Figure 3-2).

Figure 3-2 A system of inequalities.

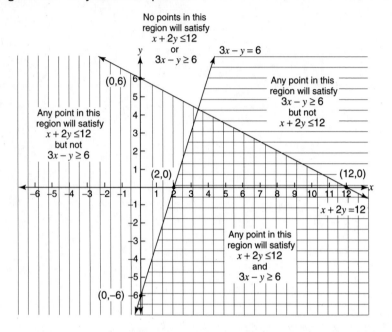

The solution to the following is, therefore, as shown in Figure 3-3.

$$x + 2y \leq 12$$
$$3x - y \geq 6$$

Example 10: Graph the solution to this system of inequalities.

$$\begin{cases} y > x + 2 \\ y < x - 1 \end{cases}$$

The graphs are shown in Figure 3-4.

Figure 3-3 The solution to the system shown in Figure 3-5.

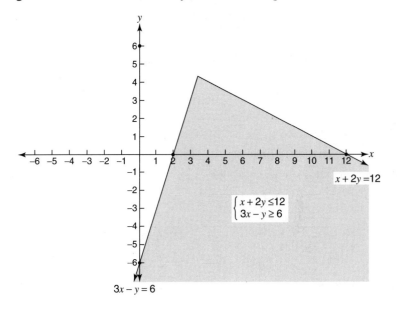

Figure 3-4 A system involving parallel lines may lack a solution.

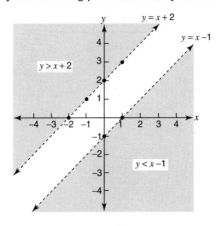

Because there are no overlapping regions, this system has no solutions.

Example 11: Solve this system by graphing.

$$\begin{cases} x \geq 2 \\ y \leq 4 \\ x - y < 6 \end{cases}$$

The graphs are shown in Figure 3-5.

Figure 3-5   Three linear equations.

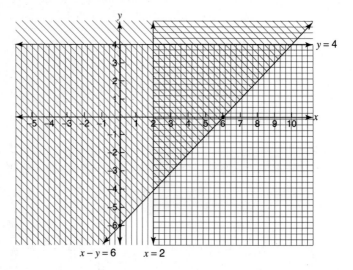

The solution is shown in Figure 3-6.

Figure 3-6 The solution triangle.

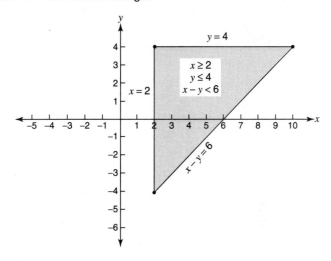

Chapter Checkout

1. Solve by substitution:

$$\begin{cases} 5x + 3y = 30 \\ 6y - 4x = 18 \end{cases}$$

2. Solve by elimination:

$$\begin{cases} 4x + 7y = 56 \\ 9x - 14y = 7 \end{cases}$$

3. Evaluate the determinant:

$$\begin{vmatrix} 4 & -8 \\ 6 & 7 \end{vmatrix}$$

4. Use Cramer's rule to solve this system:

$$\begin{cases} 2x + 3y = 39 \\ 8x - 5y = 3 \end{cases}$$

Answers: 1. $x = 3, y = 5$ 2. $x = 7, y = 4$ 3. 76   4. $x = 6, y = 9$

Chapter 4

LINEAR EQUATIONS IN THREE VARIABLES

Chapter Check-In

❑ Solving systems of equations by elimination

❑ Using matrices to solve systems of equations

❑ Using determinants to solve equation systems

Systems of equations with three variables are only slightly more complicated to solve than those with two variables (covered in the preceding chapter). The two most straightforward methods of solving these types of equations are by elimination and by using 3 x 3 matrices. Both are treated extensively in this chapter. This chapter also explores Cramer's rule — the application of determinant solution to equations.

Linear Equations: Solutions Using Elimination

To use elimination to solve a system of three equations with three variables, follow this procedure:

1. **Write all the equations in standard form cleared of decimals or fractions.**
2. **Choose a variable to eliminate; then choose any two of the three equations and eliminate the chosen variable.**
3. **Select a different set of two equations and eliminate the same variable as in Step 2.**
4. **Solve the two equations from Steps 2 and 3 for the two variables they contain.**

5. **Substitute the answers from Step 4 into any equation involving the remaining variable.**

6. **Check the solution with all three original equations.**

Example 1: Solve this system of equations using elimination.

$$\begin{cases} 4x - 2y + 3z = 1 & (1) \\ x + 3y - 4z = -7 & (2) \\ 3x + y + 2z = 5 & (3) \end{cases}$$

All the equations are already in the required form.

Choose a variable to eliminate, say x, and select two equations with which to eliminate it, say equations (1) and (2).

$$4x - 2y + 3z = 1 \xrightarrow{\text{multiply(1)}} 4x - 2y + 3z = 1$$

$$x + 3y - 4z = -7 \xrightarrow{\text{multiply(-4)}} \underline{-4x - 12y + 16z = 28}$$

$$-14y + 19z = 29 \quad (4)$$

Select a different set of two equations, say equations (2) and (3), and eliminate the same variable.

$$x + 3y - 4z = -7 \xrightarrow{\text{multiply(-3)}} -3x - 9y + 12z = 21$$

$$3x + y + 2z = 5 \xrightarrow{\text{multiply(1)}} \underline{3x + y + 2z = 5}$$

$$-8y + 14z = 26 \quad (5)$$

Solve the system created by equations (4) and (5).

$$-14y + 19z = 29 \xrightarrow{\text{multiply(-8)}} 112y - 152z = -232$$

$$-8y + 14z = 26 \xrightarrow{\text{multiply(14)}} \underline{-112y + 196z = 364}$$

$$44z = 132$$

$$z = 3$$

Now, substitute $z = 3$ into equation (4) to find y.

$$-14y + 19z = 29$$
$$-14y + 19(3) = 29$$
$$-14y + 57 = 29$$
$$-14y = -28$$
$$y = 2$$

Use the answers from Step 4 and substitute into any equation involving the remaining variable.

Using equation (2),

$$x + 3y - 4z = -7$$
$$x + 3(2) - 4(3) = -7$$
$$x + 6 - 12 = -7$$
$$x = -1$$

Check the solution in all three original equations.

$$4x - 2y + 3z = 1 \qquad\qquad x + 3y - 4z = -7$$
$$4(-1) - 2(2) + 3(3) \stackrel{?}{=} 1 \qquad\qquad -1 + 3(2) - 4(3) \stackrel{?}{=} -7$$
$$-4 - 4 + 9 \stackrel{?}{=} 1 \qquad\qquad -1 + 6 - 12 \stackrel{?}{=} -7$$
$$1 = 1 \checkmark \qquad\qquad -7 = -7 \checkmark$$

$$3x + y + 2z = 5$$
$$3(-1) + 2 + 2(3) = 5$$
$$-3 + 2 + 6 = 5$$
$$5 = 5 \checkmark$$

The solution is $x = -1$, $y = 2$, $z = 3$.

Example 2: Solve this system of equations using the elimination method.

$$\begin{cases} x = 3z - 5 & (1) \\ 2x + 2z = y + 16 & (2) \\ 7x - 5z = 3y + 19 & (3) \end{cases}$$

Write all equations in standard form.

$$x \qquad - 3z = -5 \qquad (1)$$
$$2x \ -y + 2z = 16 \qquad (2)$$
$$7x - 3y - 5z = 19 \qquad (3)$$

Notice that equation (1) already has the y eliminated. Therefore, use equations (2) and (3) to eliminate y. Then use this result, together with equation (1), to solve for x and z. Use these results and substitute into either equation (2) or (3) to find y.

$$2x - y + 2z = 16 \xrightarrow{\text{multiply}(-3)} \quad -6x + 3y - 6z = -48$$
$$7x - 3y - 5z = 19 \xrightarrow{\text{multiply}(1)} \quad \underline{7x - 3y - 5z = \quad 19}$$
$$x \qquad - 11z = -29$$

$$x - 3z = -5 \xrightarrow{\text{multiply}(-1)} \quad -x + 3z = \quad 5$$
$$x - 11z = -29 \xrightarrow{\text{multiply}(1)} \quad \underline{x - 11z = -29}$$
$$-8z = -24$$
$$z = 3$$

Substitute $z = 3$ into equation (1).

$$x - 3z = -5$$
$$x - 3(3) = -5$$
$$x - 9 = -5$$
$$x = 4$$

Substitute $x = 4$ and $z = 3$ into equation (2).

$$2x - y + 2z = 16$$
$$2(4) - y + 2(3) = 16$$
$$8 - y + 6 = 16$$
$$-y = 2$$
$$y = -2$$

Check the solution (the check is left to you).

The solution is $x = 4$, $y = -2$, $z = 3$.

Linear Equations: Solutions Using Matrices

As explained in the previous chapter, solving a system of equations by using matrices is merely an organized manner of using the elimination method.

Example 3: Solve this system of equations by using matrices.

$$\begin{cases} 2x + y - 3z = -4 \\ 4x - 2y + z = 9 \\ 3x + 5y - 2z = 5 \end{cases}$$

The goal is to arrive at a matrix of the following form.

$$\begin{bmatrix} a & b & c & \vdots & d \\ 0 & e & f & \vdots & g \\ 0 & 0 & h & \vdots & i \end{bmatrix}$$

To do this, you use row multiplications, row additions, or row switching, as shown in the following.

Put the equation in matrix form.

$$\left. \begin{array}{l} 2x + y - 3z = -4 \\ 4x - 2y + z = 9 \\ 3x + 5y - 2z = 5 \end{array} \right\} \longrightarrow \begin{bmatrix} 2 & 1 & -3 & \vdots & -4 \\ 4 & -2 & 1 & \vdots & 9 \\ 3 & 5 & -2 & \vdots & 5 \end{bmatrix} \begin{array}{l} (1) \\ (2) \\ (3) \end{array}$$

Eliminate the x-coefficient.

$$\begin{array}{r} \text{Retain row (1)} \\ \text{Add } -2 \text{ times row (1) plus 1 times row (2)} \\ \text{Add } -3 \text{ times row (1) plus 2 times row (3)} \end{array} \begin{bmatrix} 2 & 1 & -3 & \vdots & -4 \\ 0 & -4 & 7 & \vdots & 17 \\ 0 & 7 & 5 & \vdots & 22 \end{bmatrix} \begin{array}{l} (4) \\ (5) \\ (6) \end{array}$$

Eliminate the *y*-coefficient.

$$\begin{array}{r}\text{Retain row (4)}\\\text{Retain row (5)}\\\text{Add 7 times row (5) plus 4 times row (6)}\end{array}\left[\begin{array}{ccc|c}2 & 1 & -3 & -4\\0 & -4 & 7 & 17\\0 & 0 & 69 & 207\end{array}\right]\begin{array}{l}(7)\\(8)\\(9)\end{array}$$

Reinserting the variables, this system is now

$$\begin{aligned}2x + y - 3z &= -4 & (7)\\-4y + 7z &= 17 & (8)\\69z &= 207 & (9)\end{aligned}$$

Equation (9) can now be solved for *z*. That result is substituted into equation (8), which is then solved for *y*. The values for *z* and *y* are then substituted into equation (7), which is then solved for *x*.

$$69z = 207$$
$$z = 3$$
$$-4y + 7z = 17$$
$$-4y + 7(3) = 17$$
$$-4y + 21 = 17$$
$$-4y = -4$$
$$y = 1$$
$$2x + y - 3z = -4$$
$$2x + 1 - 3(3) = -4$$
$$2x + 1 - 9 = -4$$
$$2x = 4$$
$$x = 2$$

The check is left to you. The solution is $x = 2$, $y = 1$, $z = 3$.

Example 4: Solve the following system of equations, using matrices.

$$\begin{cases} 4x + 9y = 8 \\ 8x + 6z = -1 \\ 6y + 6z = -1 \end{cases}$$

Put the equations in matrix form.

$$\left.\begin{array}{r} 4x + 9y = 8 \\ 8x + 6z = -1 \\ 6y + 6z = -1 \end{array}\right\} \longrightarrow \left[\begin{array}{ccc|c} 4 & 9 & 0 & 8 \\ 8 & 0 & 6 & -1 \\ 0 & 6 & 6 & -1 \end{array}\right] \begin{array}{l}(1)\\(2)\\(3)\end{array}$$

Eliminate the x-coefficient.

$$\begin{array}{r} \text{Retain row (1)} \\ \text{Replace row (2) with row (3)} \\ \text{Add } -2 \text{ times row (1) plus 1 times row (2)} \end{array} \left[\begin{array}{ccc|c} 4 & 9 & 0 & 8 \\ 0 & 6 & 6 & -1 \\ 0 & -18 & 6 & -17 \end{array}\right] \begin{array}{l}(4)\\(5)\\(6)\end{array}$$

Eliminate the y-coefficient.

$$\begin{array}{r} \text{Retain row (4)} \\ \text{Retain row (5)} \\ \text{Add 3 times row (5) plus 1 times row (6)} \end{array} \left[\begin{array}{ccc|c} 4 & 9 & 0 & 8 \\ 0 & 6 & 6 & -1 \\ 0 & 0 & 24 & -20 \end{array}\right] \begin{array}{l}(7)\\(8)\\(9)\end{array}$$

Reinserting the variables, the system is now

$$\begin{array}{rl} 4x + 9y = 8 & (7) \\ 6y + 6z = -1 & (8) \\ 24x = -20 & (9) \end{array}$$

Equation (9) can be solved for z.

$$24z = -20$$

$$z = -\frac{20}{24}$$

$$z = -\frac{5}{6}$$

Substitute $z = -\frac{5}{6}$ into equation (8) and solve for y.

$$6y + 6z = -1$$
$$6y + 6\left(-\frac{5}{6}\right) = -1$$
$$6y - 5 = -1$$
$$6y = 4$$
$$y = \frac{4}{6}$$
$$y = \frac{2}{3}$$

Substitute $y = \frac{2}{3}$ and $z = -\frac{5}{6}$ into equation (7) and solve for x.

$$4x + 9y = 8$$
$$4x + 9\left(\frac{2}{3}\right) = 8$$
$$4x + 6 = 8$$
$$4x = 2$$
$$x = \frac{2}{4}$$
$$x = \frac{1}{2}$$

The check of the solution is left to you. The solution is $x = \frac{1}{2}$, $y = \frac{2}{3}$, $z = -\frac{5}{6}$.

Linear Equations: Solutions Using Determinants

As explained in Chapter 3, the determinant of a 2×2 matrix is defined as follows:

$$\begin{vmatrix} a & c \\ b & d \end{vmatrix} = ad - bc$$

The determinant of a 3×3 matrix can be defined as shown in the following.

$$\begin{vmatrix} a_1 & b_1 & c_1 \\ a_2 & b_2 & c_2 \\ a_3 & b_3 & c_3 \end{vmatrix} = a_1 \overbrace{\begin{vmatrix} b_2 & c_2 \\ b_3 & c_3 \end{vmatrix}}^{\text{minor determinants}} - a_2 \begin{vmatrix} b_1 & c_1 \\ b_3 & c_3 \end{vmatrix} + a_3 \begin{vmatrix} b_1 & c_1 \\ b_2 & c_2 \end{vmatrix}$$

subtract   add

$$= a_1 (b_2 c_3 - b_3 c_2) - a_2 (b_1 c_3 - b_3 c_1) + a_3 (b_1 c_2 - b_2 c_1)$$

Each minor determinant is obtained by crossing out the first column and one row.

$$\begin{vmatrix} a_1 & b_1 & c_1 \\ a_2 & b_2 & c_2 \\ a_3 & b_3 & c_3 \end{vmatrix} \quad \begin{vmatrix} a_1 & b_1 & c_1 \\ a_2 & b_2 & c_2 \\ a_3 & b_3 & c_3 \end{vmatrix} \quad \begin{vmatrix} a_1 & b_1 & c_1 \\ a_2 & b_2 & c_2 \\ a_3 & b_3 & c_3 \end{vmatrix}$$

Example 5: Evaluate the following determinant.

$$\begin{vmatrix} -2 & 4 & 1 \\ -3 & 6 & -2 \\ 4 & 0 & 5 \end{vmatrix}$$

First find the minor determinants.

$$\begin{vmatrix} -2 & 4 & 1 \\ -3 & 6 & -2 \\ 4 & 0 & 5 \end{vmatrix} \quad \begin{vmatrix} -2 & 4 & 1 \\ -3 & 6 & -2 \\ 4 & 0 & 5 \end{vmatrix} \quad \begin{vmatrix} -2 & 4 & 1 \\ -3 & 6 & -2 \\ 4 & 0 & 5 \end{vmatrix}$$

$$-2 \begin{vmatrix} 6 & -2 \\ 0 & 5 \end{vmatrix} \quad -3 \begin{vmatrix} 4 & 1 \\ 0 & 5 \end{vmatrix} \quad 4 \begin{vmatrix} 4 & 1 \\ 6 & -2 \end{vmatrix}$$

subtract   add

$$\begin{array}{ccccc} -2(30-0) & - & -3(20-0) & + & 4(-8-6) \\ -60 & - & -60 & + & -56 \\ -60 & & +60 & & -56 \quad = \quad -56 \end{array}$$

The solution is

$$\begin{vmatrix} -2 & 4 & 1 \\ -3 & 6 & -2 \\ 4 & 0 & 5 \end{vmatrix} = -56$$

To use determinants to solve a system of three equations with three variables (Cramer's rule), say x, y, and z, four determinants must be formed following this procedure:

1. **Write all equations in standard form.**
2. **Create the denominator determinant, D, by using the coefficients of x, y, and z from the equations, and evaluate it.**
3. **Create the x-numerator determinant, D_x, the y-numerator determinant, D_y, and the z-numerator determinant, D_z, by replacing the respective x, y, and z coefficients with the constants from the equations in standard form and evaluate each determinant.**

    The answers for x, y, and z are

$$x = \frac{D_x}{D}, y = \frac{D_y}{D}, z = \frac{D_z}{D}$$

Example 6: Solve this system of equations, using Cramer's rule.

$$\begin{cases} 3x + 2y - z = 2 \\ 2x - y - 3z = 13 \\ x + 3y - 2z = 1 \end{cases}$$

Find the minor determinants.

x-coefficients

y-coefficients

z-coefficients

$$D = \begin{vmatrix} 3 & 2 & -1 \\ 2 & -1 & -3 \\ 1 & 3 & -2 \end{vmatrix} = 3 \begin{vmatrix} -1 & -3 \\ 3 & -2 \end{vmatrix} - 2 \begin{vmatrix} 2 & -1 \\ 3 & -2 \end{vmatrix} + 1 \begin{vmatrix} 2 & -1 \\ -1 & -3 \end{vmatrix}$$

$$= 3[2-(-9)] - 2[-4-(-3)] + 1(-6-1)$$

$$= 3(11) \quad - \quad 2(-1) \quad + \quad 1(-7)$$

$$= \quad 33 \quad + \quad 2 \quad - \quad 7 \quad = 28$$

Use the constants to replace the x-coefficients.

constants
replacing the
x-coefficients

$$D_x = \begin{vmatrix} 2 & 2 & -1 \\ 13 & -1 & -3 \\ 1 & 3 & -2 \end{vmatrix} = 2 \begin{vmatrix} -1 & -3 \\ 3 & -2 \end{vmatrix} - 13 \begin{vmatrix} 2 & -1 \\ 3 & -2 \end{vmatrix} + 1 \begin{vmatrix} 2 & -1 \\ -1 & -3 \end{vmatrix}$$

$$= 2[2-(-9) - 13[-4-(-3)] + 1(-6-1)]$$

$$= \quad 2(11) \quad - \quad 13(-1) \quad + \quad 1(-7)$$

$$= \quad 22 \quad + \quad 13 \quad - \quad 7 \quad = 28$$

Use the constants to replace the y-coefficients.

constants
replacing the
y-coefficients
↓

$$D_y = \begin{vmatrix} 3 & 2 & -1 \\ 2 & 13 & -3 \\ 1 & 1 & -2 \end{vmatrix} = 3\begin{vmatrix} 13 & -3 \\ 1 & -2 \end{vmatrix} - 2\begin{vmatrix} 2 & -1 \\ 1 & -2 \end{vmatrix} + 1\begin{vmatrix} 2 & -1 \\ 13 & -3 \end{vmatrix}$$

$$= 3[-26-(-3)] - 2[-4-(-1)] + 1[-6-(-13)]$$

$$= 3(-23) \quad - \quad 2(-3) \quad + \quad 1(7)$$

$$= \quad -69 \quad + \quad 6 \quad + \quad 7 \quad = -56$$

Use the constants to replace the z-coefficients.

constants
replacing the
z-coefficients
↓

$$D_z = \begin{vmatrix} 3 & 2 & 2 \\ 2 & -1 & 13 \\ 1 & 3 & 1 \end{vmatrix} = 3\begin{vmatrix} -1 & 13 \\ 3 & 1 \end{vmatrix} - 2\begin{vmatrix} 2 & 2 \\ 3 & 1 \end{vmatrix} + 1\begin{vmatrix} 2 & 2 \\ -1 & 13 \end{vmatrix}$$

$$= 3(-1-39) - 2(2-6) + 1[26-(-2)]$$

$$= 3(-40) \quad - \quad 2(-4) \quad + \quad 1(28)$$

$$= \quad -120 \quad + \quad 8 \quad + \quad 28 \quad = -84$$

Therefore,

$$x = \frac{D_x}{D} = \frac{28}{28} = 1, \; y = \frac{D_y}{D} = -\frac{56}{28} = -2, \; z = \frac{D_z}{D} = -\frac{84}{28} = -3$$

The check is left to you. The solution is $x = 1$, $y = -2$, $z = -3$.

If the denominator determinant, D, has a value of zero, then the system is either inconsistent or dependent. The system is dependent if all the determinants have a value of zero. The system is inconsistent if at least one of the determinants, D_x, D_y, or D_z, has a value not equal to zero and the denominator determinant has a value of zero.

Chapter Checkout

1. Solve this system of equations by the method of elimination.

$$\begin{cases} 5x - 2y + 3z = 22 \\ 2x + 3y - 4z = -2 \\ 3x + 4y + z = 30 \end{cases}$$

2. Solve this system of equations by the method of elimination.

$$\begin{cases} x + 2y + 20 = 3z \\ y + 2z + 3x = -1 \\ 3z + 2x - 2y = 14 \end{cases}$$

3. Use Cramer's rule to solve this system of equations.

$$\begin{cases} 2x + 3y - z = -7 \\ x - 2y + z = -2 \\ 3x + y + 2z = -7 \end{cases}$$

Answers: 1. $x = 3$, $y = 4$, $z = 5$ 2. $x = -2$, $y = -3$, $z = 4$
3. $x = -3$, $y = 0$, $z = 1$

Chapter 5

POLYNOMIAL ARITHMETIC

Chapter Check-In

❏ Adding and subtracting polynomials

❏ Multiplying polynomials

❏ Working with special products of binomials

❏ Dividing binomials

❏ Using synthetic division

Polynomials are expressions containing more than one term, with each term separated from the preceding one by a plus or minus sign. There is no maximum length to a polynomial. Some arithmetic operations with polynomials need just common sense, while others require special techniques. All are treated within this chapter.

Adding and Subtracting Polynomials

In order to add and subtract polynomials successfully, you must understand what monomials, binomials, and trinomials are; what constitutes "like terms"; and the difference between ascending and descending order.

Monomial, binomial, and trinomial

A **monomial** is an expression that could be a numeral, a variable, or the product of numerals and variables. If the expression has variables, certain restrictions apply to make it a monomial.

■ Variables must have whole number exponents.

■ Variables do not appear under simplified radical expressions.

■ Denominators do not contain variables.

The following expressions are examples of monomials.

$$-12, \ a, \ 3t^2, \ \tfrac{5}{8}x^2y^3, \ \frac{2(x+y^2)}{3}$$

The following are expressions that are *not* monomials.

x^{-2} (The variable has an exponent that isn't a whole number.)

$\sqrt{x}$ (A variable is under a simplified radical.)

$\dfrac{3(x+2)}{x+3}$ (The denominator has a variable.)

A **binomial** is an expression that is the sum of two monomials.

A **trinomial** is an expression that is the sum of three monomials.

A **polynomial** is an expression that is a monomial or the sum of two or more monomials.

Like terms or similar terms

Two or more monomials with identical variable expressions are called **like terms** or **similar terms.** The following are like terms, since their variable expressions are all x^2y:

$$5x^2y, \ -3x^2y, \ \tfrac{2}{7}x^2y$$

These are not like terms, since their variable expressions are not all the same:

$$-5x^2y^2, \ 4x^2y, \ \tfrac{1}{2}xy^2$$

In order to add monomials, they must be like terms. *Unlike terms cannot be added together.* To add like terms, follow this procedure.

1. **Add their numerical coefficients.**
2. **Keep the variable expression.**

Example 1: Find the following sums.

    (a) $4x^2y + 8x^2y$

    (b) $-9abc + 3abc$

    (c) $9xy + 7x - 28xy - 4x$

    (a) $12x^2y$ (b) $-6abc$ (c) $-19xy + 3x$

Note that in answer (c), because $-19xy$ and $3x$ are unlike terms, they cannot be added together.

Ascending and descending order

When working with polynomials that involve only one variable, the general practice is to write them so that the exponents on the variable *decrease* from left to right. The polynomial is then said to be written in **descending order.**

When a polynomial in one variable is written so that the exponents *increase* from left to right, it is referred to as being written in **ascending order.**

Example 2: Rewrite the following polynomial in descending powers of x.

$$4y^4 + 12 - 15x^2 + 13x^3y + 17xy^2$$
$$13x^3y - 15x^2 + 17xy^2 + 4y^4 + 12$$

To add two or more polynomials, add like terms and arrange the answer in descending (or ascending if asked) powers of one variable.

Example 3: Find the following sum.

$$(x^2 + x^3 - 3x) + (4 - 5x^2 + 3x^3) + (10 - 8x^2 - 5x)$$
$$(x^3 + 3x^3) + (x^2 - 5x^2 - 8x^2) + (-3x - 5x) + (4 + 10)$$
$$= 4x^3 - 12x^2 - 8x + 14$$

This problem can also be added vertically. First rewrite each polynomial in descending order, one above the other, placing like terms in the same column.

$$
\begin{array}{r}
x^3 + x^2 - 3x \\
3x^3 - 5x^2 + 4 \\
-8x^2 - 5x + 10\\
\hline
4x^3 - 12x^2 - 8x + 14
\end{array}
$$

To subtract one polynomial from another, add its opposite.

Example 4: Subtract $(4x^2 - 7x + 3)$ from $(6x^2 + 4x - 9)$.

Done horizontally,

$$(6x^2 + 4x - 9) - (4x^2 - 7x + 3)$$
$$= 6x^2 + 4x - 9 - 4x^2 + 7x - 3$$
$$= 2x^2 + 11x - 12$$

Done vertically,

$$
\begin{array}{r}
6x^2 + 4x - 9 \quad \rightarrow \quad 6x^2 + 4x - 9\\
-(4x^2 - 7x + 3) \quad \rightarrow \quad -4x^2 + 7x - 3\\
\hline
2x^2 + 11x - 12
\end{array}
$$

Multiplying Polynomials

The following are rules regarding the multiplying of variable expressions.

- **Rule 1:** To multiply monomials with the same base, keep the base and add the powers:

$$x^a x^b = x^{a+b}$$

- **Rule 2:** To find the power of a base, keep the base and multiply the powers.

$$(x^a)^b = x^{ab}$$

- **Rule 3:** To find a power of a product, raise each factor in the product to the power.

$$(xy)^a = x^a y^a$$

Example 5: Simplify each of the following multiplication problems and state which rule above was applied.

(a) yy^5 (b) $(x^4)^3$ (c) $(-2x^4 y^2 z^3)^5$ (d) $a^3(a^2 b^3)^4$

(a) $yy^5 = y^{1+5} = y^6$ (Rule 1)

(b) $(x^4)^3 = x^{4 \times 3} = x^{12}$ (Rule 2)

(c) $(-2x^4 y^2 z^3)^5 = (-2)^5 x^{4 \times 5} y^{2 \times 5} z^{3 \times 5} = -32 x^{20} y^{10} z^{15}$ (Rule 3 and Rule 2)

(d) $a^3(a^2 b^3)^4 = a^3(a^2)^4(b^3)^4$ (Rule 3)

$= a^3 a^8 b^{12}$ (Rule 2)

$= a^{11} b^{12}$ (Rule 1)

To multiply monomials together, follow this procedure.

1. **Multiply the numerical coefficients together.**
2. **Multiply the variables together.**
3. **Multiply the results together.**

Example 6: Simplify each of the following.

(a) $(4x^2)(3x^3)$ (b) $(-8a^3 b^2)(2a^2 b^2)^3$

(a) $(4x^2)(3x^3) = (4 \cdot 3)(x^2 x^3) = 12x^5$

(b) $(-8a^3 b^2)(2a^2 b^2)^3 = (-8a^3 b^2)(8a^6 b^6) = -64a^9 b^8$

To multiply polynomials together, multiply each term in one polynomial by each term in the other polynomial. Then simplify if possible.

Example 7: Multiply each of the following.

$$\text{(a) } 5x(3x^2 - 4x + 2)$$
$$\text{(b) } (4x - 2)(3x + 5)$$
$$\text{(c) } (x + y)(x^2 - xy + y^2)$$

The following shows how each equation is multiplied both horizontally and vertically.

Equation (a) done horizontally.

$$5x(3x^2 - 4x + 2) = 15x^3 - 20x^2 + 10x$$

Equation (a) done vertically.

$$
\begin{array}{r}
3x^2 - 4x + 2 \\
\times \qquad \quad 5x \\
\hline
15x^3 - 20x^2 + 10x
\end{array}
$$

Equation (b) done horizontally.

$$(4x - 2)(3x + 5) = 12x^2 + 20x - 6x - 10$$
$$= 12x^2 + 14x - 10$$

Equation (b) done vertically.

$$
\begin{array}{r}
3x + 5 \\
4x - 2 \\
\hline
\times \qquad \\
-6x - 10 \\
12x^2 + 20x \\
\hline
12x^2 + 14x - 10
\end{array}
$$

Equation (c) done horizontally.

$$(x + y)(x^2 - xy + y^2) = x^3 - x^2y + xy^2 + x^2y - xy^2 + y^3$$

$$= x^3 + y^3$$

Equation (c) done vertically.

$$
\begin{array}{r}
x^2 - xy + y^2 \\
\times \quad x + y \\
\hline
x^2y - xy^2 + y^3 \\
x^3 - x^2y + xy^2 \\
\hline
x^3 \qquad\qquad + y^3
\end{array}
$$

Special Products of Binomials

Two binomials with the same two terms but opposite signs separating the terms are called **conjugates** of each other. Following are examples of conjugates:

$$3x + 2 \qquad \text{and} \qquad 3x - 2$$
$$-5a - 4b \qquad \text{and} \qquad -5a + 4b$$

Example 8: Find the product of the following conjugates.

(a) $(3x + 2)(3x - 2)$ (b) $(-5a - 4b)(-5a + 4b)$

(a) $(3x + 2)(3x - 2)$
$$= 9x^2 - 6x + 6x - 4$$
$$= 9x^2 - 4$$

(b) $(-5a - 4b)(-5a + 4b)$
$$= 25a^2 - 20ab + 20ab - 16b^2$$
$$= 25a^2 - 16b^2$$

Notice that when conjugates are multiplied together, the answer is the difference of the squares of the terms in the original binomials.

The product of conjugates produces a special pattern referred to as a **difference of squares.** In general,

$$(x + y)(x - y) = x^2 - y^2$$

The squaring of a binomial also produces a special pattern.

Example 9: Simplify each of the following.

(a) $(4x + 3)^2$ (b) $(6a - 7b)^2$

(a) $(4x + 3)^2 = (4x + 3)(4x + 3)$

$$= 16x^2 + 12x + 12x + 9$$
$$= 16x^2 + 2(12x) + 9$$
$$= 16x^2 + 24x + 9$$

(b) $(6a - 7b)^2 = (6a - 7b)(6a - 7b)$

$$= 36a^2 - 42ab - 42ab + 49b^2$$
$$= 36a^2 - 2(42ab) + 49b^2$$
$$= 36a^2 - 84ab + 49b^2$$

First, notice that the answers are trinomials. Second, notice that there is a pattern in the terms:

- The first and last terms are the squares of the first and last terms of the binomial.

- The middle term is *twice* the product of the two terms in the binomial.

The pattern produced by squaring a binomial is referred to as a **square trinomial.** In general,

$$(x + y)^2 = x^2 + 2xy + y^2$$

and

$$(x - y)^2 = x^2 - 2xy + y^2$$

Example 10: Do the following special binomial products mentally.

(a) $(3x + 4y)^2$ (b) $(6x + 11)(6x - 11)$

(a) $(3x + 4y)^2 = 9x^2 + 24xy + 16y^2$

(b) $(6x + 11)(6x - 11) = 36x^2 - 121$

Dividing Polynomials

Example 11: Simplify each of the following division expressions and find the pattern involving the exponents.

$$\text{(a) } \frac{a^8}{a^2} \qquad \text{(b) } \frac{a^8}{a^3}$$

(a)
$$\frac{a^8}{a^2} = \frac{\overset{1\cdot 1}{\cancel{a}\,\cancel{a}\,aaaaaa}}{\underset{1\cdot 1}{\cancel{a}\,\cancel{a}}} = a^6$$

(b)
$$\frac{a^8}{a^3} = \frac{\overset{1\cdot 1\cdot 1}{\cancel{a}\,\cancel{a}\,\cancel{a}\,aaaaa}}{\underset{1\cdot 1\cdot 1}{\cancel{a}\,\cancel{a}\,\cancel{a}}} = a^5$$

These examples suggest the following rule.

$$\frac{a^m}{a^n} = a^{m-n} \qquad \text{as long as a} \neq 0$$

Example 12: Simplify each of the following.

$$\text{(a) } \frac{a^5}{a^5} \qquad \text{(b) } \frac{a^5}{a^6} \qquad \text{(c) } \frac{a^5}{a^8}$$

(a) $\quad \dfrac{a^5}{a^5} = a^{5-5} = a^0 = 1$

(b) $\quad \dfrac{a^5}{a^6} = a^{5-6} = a^{-1}$ also $\dfrac{a^5}{a^6} = \dfrac{\overset{1\cdot 1\cdot 1\cdot 1\cdot 1}{\cancel{a}\,\cancel{a}\,\cancel{a}\,\cancel{a}\,\cancel{a}}}{\underset{1\cdot 1\cdot 1\cdot 1\cdot 1}{\cancel{a}\,\cancel{a}\,\cancel{a}\,\cancel{a}\,\cancel{a}\, a}} = \dfrac{1}{a} \qquad$ so $a^{-1} = \dfrac{1}{a}$

(c) $\quad \dfrac{a^5}{a^8} = a^{5-8} = a^{-3}$ also $\dfrac{a^5}{a^8} = \dfrac{\overset{1\cdot 1\cdot 1\cdot 1\cdot 1}{\cancel{a}\,\cancel{a}\,\cancel{a}\,\cancel{a}\,\cancel{a}}}{\underset{1\cdot 1\cdot 1\cdot 1\cdot 1}{\cancel{a}\,\cancel{a}\,\cancel{a}\,\cancel{a}\,\cancel{a}\, aaa}} = \dfrac{1}{a^3} \qquad$ so $a^{-3} = \dfrac{1}{a^3}$

These examples suggest the following rule:

$$a^{-n} = \frac{1}{a^n} \text{ and } \frac{1}{a^{-n}} = a^n \qquad \text{for all } (a \neq 0)$$

Generally, when simplifying expressions, write the final result *without* the use of negative exponents.

Example 13: Simplify each of the following.

$$\text{(a) } \frac{x^{2a}}{x^{3a}} \qquad \text{(b) } \frac{3^{4x-3}}{3^{2x-4}}$$

(a) $\dfrac{x^{2a}}{x^{3a}} = x^{2a-3a} = x^{-a} = \dfrac{1}{x^{a}}$

(b) $\dfrac{3^{4x-3}}{3^{2x-4}} = 3^{(4x-3)-(2x-4)} = 3^{4x-3-2x+4} = 3^{2x+1}$

To divide a monomial by another monomial, follow the procedure.

1. **Divide the numerical coefficients.**
2. **Divide the variables.**
3. **Multiply the results together.**

Example 14: Simplify the following expressions by dividing correctly.

$$\text{(a) } \frac{12x^8}{4x^3} \qquad \text{(b) } \frac{-27w^3 t^7}{-3w^3 t^{12}} \qquad \text{(c) } \frac{-15x^3 y^4}{10y^5 z}$$

(a) $\dfrac{12x^8}{4x^3} = \dfrac{12}{4} \cdot \dfrac{x^8}{x^3} = 3x^5$

(b) $\dfrac{-27w^3 t^7}{-3w^3 t^{12}} = \dfrac{-27}{-3} \cdot \dfrac{w^3}{w^3} \cdot \dfrac{t^7}{t^{12}} = 9 \cdot 1 \cdot \dfrac{1}{t^5} = \dfrac{9}{t^5}$

(c) $\dfrac{-15x^3 y^4}{10y^5 z} = \dfrac{-15}{10} \cdot \dfrac{x^3}{1} \cdot \dfrac{y^4}{y^5} \cdot \dfrac{1}{z} = \dfrac{-3x^3}{2yz}$

To divide a polynomial by a monomial, divide each term of the polynomial by the monomial.

Example 15: Simplify the given expression by dividing correctly.

$$\frac{15r^2 s - 7rs^2 + 6s^3}{-3r^2}$$

$$\frac{15r^2 s - 7rs^2 + 6s^3}{-3r^2} = \frac{15r^2 s}{-3r^2} - \frac{7rs^2}{-3r^2} + \frac{6s^3}{-3r^2}$$

$$= -5s + \frac{7s^2}{3r} - \frac{2s^3}{r^2}$$

To divide a polynomial by a polynomial, a procedure similar to long division in arithmetic is used. The procedure calls for four steps: divide, multiply, subtract, bring down. This procedure is repeated until there is no value to bring down.

Example 16: Divide $x^2 + 3x^3 - 5$ by $4 + x$.

First, arrange both polynomials in descending order, leaving space or filling in a place holder for any missing terms.

$$x + 4 \overline{\smash{)}\,3x^3 + x^2 + 0x - 5}$$
$$\uparrow$$

place holder, since there
is no "x" term

Now *divide* $3x^3$ by x and bring this to the top as the first part of the answer.

$$\frac{3x^3}{x} = 3x^2 \longrightarrow \begin{array}{r} 3x^2 \\ x + 4 \overline{\smash{)}\,3x^3 \quad + x^2 + 0x - 5} \end{array}$$

Multiply $3x^2$ by the divisor $x + 4$ and place these in the columns with like terms.

$$\begin{array}{r} 3x^2 \\ x + 4 \overline{\smash{)}\,3x^3 + \quad x^2 + 0x - 5} \end{array}$$
$$3x^2(x + 4) = 3x^3 + 12x^2 \longrightarrow \quad \underline{3x^3 + 12x^2}$$

Subtract. Remember, to subtract is to add the opposite.

$$\begin{array}{r} 3x^2 \\ x + 4 \overline{\smash{)}\,3x^3 + \quad x^2 + 0x - 5} \\ \text{subtract} \rightarrow \quad -\left(3x^3 + 12x^2\right) \\ \hline -11x^2 \end{array}$$

Bring down the next term and start the procedure again.

$$\begin{array}{r} 3x^2 \\ x+4 \overline{\smash{\big)}\ 3x^3 + x^2 + 0x - 5} \\ \underline{-\left(3x^3 + 12x^2\right)} \quad \downarrow \\ -11x^2 + 0x \end{array}$$

Divide. $\dfrac{-11x^2}{x} = -11x$

$$\begin{array}{r} 3x^2 - 11x \\ x+4 \overline{\smash{\big)}\ 3x^3 + x^2 + 0x - 5} \\ \underline{-\left(3x^3 + 12x^2\right)} \\ -11x^2 + 0x \end{array}$$

Multiply and subtract.

$-11x(x+4) = -11x^2 - 44x \longrightarrow \underline{-\left(11x^2 + 44x\right)}$

Bring down.
$$44x - 5$$

Since a value was brought down, start the process over again.

Divide. $\dfrac{44x}{x} = 44$

$$\begin{array}{r} 3x^2 - 11x + 44 \\ x+4 \overline{\smash{\big)}\ 3x^3 + x^2 + 0x - 5} \\ \underline{-\left(3x^3 + 12x^2\right)} \\ -11x^2 + 0x \\ \underline{-\left(11x^2 + 44x\right)} \end{array}$$

$$44x - 5$$

Multiply and subtract.

$44(x+4) = 44x + 176 \longrightarrow \underline{-(44x + 176)}$
$$-181$$

At this point, there is no term to bring down. The −181 is the remainder. As in arithmetic, remainders are written over the divisor. So the final answer to the division problem is

$$3x^2 - 11x + 44 - \frac{181}{x+4}$$

Example 17: Divide $64x^3 - 27$ by $4x - 3$.

$$\frac{64x^3}{4x} = 16x^2$$

$16x^2(4x-3) = 64x^3 - 48x^2$

$$\frac{48x^2}{4x} = 12x$$

$12x(4x-3) = 48x^2 - 36x$

$$\frac{36x}{4x} = 9$$

$9(4x-3) = 36x - 27$

$$
\begin{array}{r}
16x^2 + 12x + 9 \\
4x-3 \overline{)\; 64x^3 + \quad 0x^2 + \quad 0x - 27} \\
\underline{-\left(64x^3 - 48x^2\right)} \\
48x^2 + \quad 0x \\
\underline{-\left(48x^2 - 36x\right)} \\
36x - 27 \\
\underline{-\left(36x - 27\right)} \\
0
\end{array}
$$

The answer is $16x^2 + 12x + 9$

Synthetic Division

Synthetic division is a shortcut for polynomial division when the divisor is of the form $x - a$. Only coefficients are used when dividing with synthetic division.

Example 18: Divide $(2x - 11 + 3x^3)$ by $(x - 3)$.

First, this problem will be done in the traditional manner. Then it will be done by using the synthetic division method.

In the traditional manner,

$$
\begin{array}{r}
3x^2 + 9x + 29 \\
x-3 \overline{)\; 3x^3 + 0x^2 + \quad 2x - 11} \\
\underline{-\left(3x^3 - 9x^2\right)} \\
9x^2 + \quad 2x \\
\underline{-\left(9x^2 - 27x\right)} \\
29x - 11 \\
\underline{-(29x - 87)} \\
76
\end{array}
$$

The answer is $3x^2 + 9x + 29 + \dfrac{76}{x-3}$

To do the problem using synthetic division, follow this procedure:

1. **Write the polynomial being divided in descending order. Then write only its coefficients, using 0 for any missing terms.**

$$3x^3 + 0x^2 + 2x - 11$$
$$\begin{array}{cccc} 3 & 0 & 2 & -11 \end{array}$$

2. **Write the constant, *a*, of the divisor, *x – a*, to the left. In this problem, *a* = 3.**

$$\underline{3|}\quad 3 \quad 0 \quad 2 \quad -11$$

3. **Bring down the first coefficient as shown.**

$$\begin{array}{c|cccc} \underline{3|} & 3 & 0 & 2 & -11 \\ & \downarrow \\ \hline & 3 \end{array}$$

4. **Multiply the first coefficient by *a*. Then write this product under the second coefficient.**

$$\begin{array}{c|cccc} & \underline{3|} & 3 & 0 & 2 & -11 \\ (3 \times 3 = 9) & \longrightarrow & & 9 \\ \hline & & 3 \end{array}$$

5. **Add the second coefficient with the product and write the sum as shown.**

$$\begin{array}{c|cccc} \underline{3|} & 3 & 0 & 2 & -11 \\ & & 9 \\ \hline & 3 & 9 \end{array}$$

6. **Continue this process of multiplying and adding until there is a sum for the last column.**

$$\begin{array}{c|cccc} \underline{3|} & 3 & 0 & 2 & -11 \\ & & 9 & 27 & 87 \\ \hline & 3 & 9 & 29 & 76 \end{array}$$

The numbers along the bottom row are the coefficients of the quotients with the powers of *x* in descending order. The last coefficient is the remainder. The first power is one less than the highest power of the polynomial that was being divided.

The division answer is $3x^2 + 9x + 29 + \dfrac{76}{x-3}$

Example 19: Divide $(5x^4 + 6x^3 - 9x^2 - 7x + 6)$ by $(x + 2)$ using the synthetic method.

The divisor, $x + 2$, is the same as $x - (-2)$. Now, the divisor is of the form $x - a$, with $a = -2$.

$$
\begin{array}{r|rrrrr}
-2 & 5 & 6 & -9 & -7 & 6 \\
   &   & -10 & 8 & 2 & 10 \\
\hline
   & 5 & -4 & -1 & -5 & 16
\end{array}
$$

The answer is $5x^3 - 4x^2 - x - 5 + \dfrac{16}{x+2}$

Chapter Checkout

1. Find the following difference:

$$\left(5x^4 + 3x^3 + 2x^2 - 4x\right) - \left(2x^4 + 5x^2 + 4x - 6\right)$$

2. Find the following sum:

$$\left(5ab^2 - 2ab + 3a^2b\right) + \left(3ab - 2ab^2 + a^2b\right)$$

3. Simplify each of the following multiplication problems:

a) $b^2 \cdot b^3$ b) $\left(y^4\right)^3$ c) $\left(-3x^4y^3z^2\right)^3$

d) $(2x - 3)(x + 4)$ e) $(3x + 2)(3x - 2)$

Answers: 1. $3x^4 + 3x^3 - 3x^2 - 8x + 6$ 2. $4a^2b + ab + 3ab^2$

3. a. b^5 b. y^{12} c. $-27x^{12}y^9z^6$ d. $2x^2 + 5x - 12$ e. $9x^2 - 4$

Chapter 6

FACTORING POLYNOMIALS

Chapter Check-In

- ❏ Factoring out the greatest common factor
- ❏ Working with the difference of squares
- ❏ Factoring the difference of cubes
- ❏ Factoring various types of trinomial
- ❏ Solving equations by factoring

To **factor** a polynomial means to rewrite the polynomial as a product of simpler polynomials or of polynomials and monomials. Because polynomials may take many different forms, there are many different techniques available for factoring them. All of the common methods are explored in this chapter, before going on to see how factoring may be used as a powerful tool for solving equations of a degree higher than one.

Greatest Common Factor

The first method of factoring is called factoring out the **GCF** (**greatest common factor**).

Example 1: Factor $5x + 5y$.

Since each term in this polynomial involves a factor of 5, then 5 is a common factor of the polynomial.

$$5x + 5y = 5(x + y)$$

$$5x \div 5 = x$$
$$5y \div 5 = y$$

Example 2: Factor $24x^3 - 16x^2 + 8x$.

x is a common factor for all three terms. Also, the numbers 24, −16, and 8 all have the common factors of 2, 4, and 8. The *greatest common factor* is $8x$.

$$24x^3 - 16x^2 + 8x = 8x(3x^2 - 2x + 1)$$

$24x^3 \div 8x = 3x^2$
$-16x^2 \div 8x = -2x$
$8x \div 8x = 1$

Difference of Squares

Recall that the product of conjugates produces a pattern called a **difference of squares.**

$$(a + b) \text{ and } (a - b) \text{ are conjugates}$$

$$(a + b)(a - b) = \underbrace{a^2 - b^2}_{\substack{\text{difference} \\ \text{of squares}}}$$

Example 3: Factor $x^2 - 16$.

This polynomial results from the subtraction of two values that are each the square of some expression.

$$x^2 = (x)^2 \text{ and } 16 = (4)^2$$

So

$$\underbrace{x^2 - 16}_{\substack{\text{difference} \\ \text{of squares}}} = (x)^2 - (4)^2$$

$$= \underbrace{(x + 4)(x - 4)}_{\text{conjugates}}$$

Example 4: Factor $25x^2y^2 - 36z^2$.

$$25x^2y^2 = (5xy)^2 \text{ and } 36z^2 = (6z)^2$$

$$25x^2y^2 - 36z^2 = (5xy)^2 - (6z)^2$$

$$= (5xy + 6z)(5xy - 6z)$$

Example 5: Factor $(a + b)^2 - (c - d)^2$.

$$(a + b)^2 - (c - d)^2 = [(a + b) + (c - d)][(a + b) - (c - d)]$$

$$= (a + b + c - d)(a + b - c + d)$$

Example 6: Factor $y^2 + 9$.

Even though y^2 and 9 are square numbers, the expression $y^2 + 9$ is *not* a *difference* of squares and is not factorable.

Many polynomials require more than one method of factoring to be completely factored into a product of polynomials. Because of this, a sequence of factoring methods must be used.

■ First, try to factor by using the GCF.

■ Second, try to factor by using the difference of squares.

Example 7: Factor $9x^2 - 36$.

$$9x^2 - 36 = 9\underbrace{\left(x^2 - 4\right)}_{\substack{\text{difference} \\ \text{of squares}}} \qquad \text{GCF of 9}$$

$$= 9(x+2)(x-2)$$

Example 8: Factor $8(x+y)^2 - 18$.

$$8\left(x+y\right)^2 - 18 = 2\underbrace{\left[4\left(x+y\right)^2 - 9\right]}_{\text{difference of squares}} \qquad \text{GCF of 2}$$

Note: $4(x+y)^2 = [2(x+y)]^2$ and $9 = 3^2$.

$$= 2[2(x+y)+3][2(x+y)-3]$$

$$= 2(2x+2y+3)(2x+2y-3)$$

Sum or Difference of Cubes

A polynomial in the form $a^3 + b^3$ is called a **sum of cubes.** A polynomial in the form $a^3 - b^3$ is called a **difference of cubes.**

Both of these polynomials have similar factored patterns:

A sum of cubes:

$$a^3 + b^3 = (a + b)(a^2 - ab + b^2)$$

same sign — opposite sign — always +

A difference of cubes:

$$a^3 - b^3 = (a - b)(a^2 + ab + b^2)$$

same sign

opposite sign

always +

Example 9: Factor $x^3 + 125$.

$$x^3 + 125 = (x)^3 + (5)^3$$
$$= (x + 5)[x^2 - (x)(5) + 5^2]$$
$$= (x + 5)(x^2 - 5x + 25)$$

Example 10: Factor $8x^3 - 27$.

$$8x^3 - 27 = (2x)^3 - (3)^3$$
$$= (2x - 3)[(2x)^2 + (2x)(3) + 3^2]$$
$$= (2x - 3)(4x^2 + 6x + 9)$$

Example 11: Factor $2x^3 + 128y^3$.

First find the GFC. GFC = 2

$$2x^3 + 128y^3 = 2(x^3 + 64y^3)$$
$$= 2[(x)^3 + (4y)^3]$$
$$= 2[x + 4y][x^2 - (x)(4y) + (4y)^2]$$
$$= 2(x + 4y)(x^2 - 4xy + 16y^2)$$

Example 12: Factor $x^6 - y^6$.

First, notice that $x^6 - y^6$ is both a difference of squares and a difference of cubes.

$$x^6 - y^6 = (x^3)^2 - (y^3)^2 \qquad x^6 - y^6 = (x^2)^3 - (y^2)^3$$

In general, factor a difference of squares before factoring a difference of cubes.

$$x^6 - y^6 = \underbrace{\left(x^3\right)^2 - \left(y^3\right)^2}_{\substack{\text{difference} \\ \text{of squares}}}$$

$$= \underbrace{\left(x^3 + y^3\right)}_{\substack{\text{sum of} \\ \text{cubes}}} \underbrace{\left(x^3 - y^3\right)}_{\substack{\text{difference} \\ \text{of cubes}}}$$

$$= \left[(x+y)(x^2 - xy + y^2) \right] \left[(x-y)(x^2 + xy + y^2) \right]$$
$$= (x+y)(x^2 - xy + y^2)(x-y)(x^2 + xy + y^2)$$

Trinomials of the Form $x^2 + bx + c$

To factor polynomials of the form $x^2 + bx + c$, begin with two pairs of parentheses with x at the left of each.

$$(x \quad)(x \quad)$$

Next, find two integers whose product is c and whose sum is b and place them at the right of the parentheses.

Example 13: Factor $x^2 + 8x + 12$.

$$x^2 + 8x + 12 = (x \quad)(x \quad)$$

12 can be factored in a variety of ways:

$(1)(12), \quad (-1)(-12), \quad (2)(6), \quad (-2)(-6), \quad (3)(4), \quad (-3)(-4)$

The only combination whose sum is also 8 is (2)(6), so

$$x^2 + 8x + 12 = (x + 2)(x + 6)$$

Example 14: Factor $x^2 - 7x - 18$.

−18 can be factored in the following ways:

$(1)(-18), \quad (-1)(18), \quad (2)(-9), \quad (-2)(9), \quad (3)(-6), \quad (-3)(6)$

The only combination whose sum is also −7 is (2)(−9), so

$$x^2 - 7x + 18 = (x + 2)(x - 9)$$

Example 15: Factor $x^2 - 6x + 9$.

9 can be factored as

$$(1)(9), \qquad (-1)(-9), \qquad (3)(3), \qquad (-3)(-3)$$

The only combination whose sum is −6 is (−3)(−3), so

$$x^2 - 6x + 9 = (x - 3)(x - 3) = (x - 3)^2$$

Trinomials of the Form $ax^2 + bx + c$

Study this pattern for multiplying two binomials:

$$(3x + 5)(4x - 3) = 12x^2 - 9x + 20x - 15$$

first   outer   inner   last

$$= 12x^2 + 11x - 15$$

first   sum   last
of outer
& inner

Example 16: Factor $2x^2 - 5x - 12$.

Begin by writing two pairs of parentheses.

$$(\quad)(\quad)$$

For the first positions, find two factors whose product is $2x^2$. For the last positions, find two factors whose product is -12. Below are the possibilities. With each possibility, the sum of outer and inner products is included.

1. $(x + 1)(\underline{2x - 12})$; $-10x$
2. $(x - 1)(\underline{2x + 12})$; $+10x$
3. $(x - 12)(2x + 1)$; $-23x$
4. $(x + 12)(2x - 1)$; $+23x$
5. $(x + 2)(\underline{2x - 6})$; $-2x$
6. $(x - 2)(\underline{2x + 6})$; $+2x$
7. $(x - 6)(\underline{2x + 2})$; $-10x$
8. $(x + 6)(\underline{2x - 2})$; $+10x$
9. $(x + 3)(\underline{2x - 4})$; $+2x$
10. $(x - 3)(\underline{2x + 4})$; $-2x$
11. $(x - 4)(\underline{2x + 3})$; $-5x$
12. $(x + 4)(\underline{2x - 3})$; $+5x$

Only possibility 11 will multiply out to produce the original polynomial. Therefore,

$$2x^2 - 5x - 12 = (x - 4)(2x + 3)$$

Because many possibilities exist, some shortcuts are advisable:

■ **Shortcut 1:** Be sure the GCF, if there is one, has been factored out.

■ **Shortcut 2:** Try factors closest to one another first. For example, when considering factors of 12, try 3 and 4 before trying 6 and 2 and try 6 and 2 before trying 1 and 12.

■ **Shortcut 3:** Avoid creating binomials that will have a GCF within them. This shortcut eliminates possibilities 1, 2, 5, 6, 7, 8, 9, and 10 (look at the underlined binomials; their terms each have some common factor), leaving only four possibilities to consider. Of the four remaining possibilities, 11 and 12 would be considered first using shortcut 2.

Example 17: Factor $8x^2 - 26x + 20$.

$$8x^2 - 26x + 20 = 2(4x^2 - 13x + 10) \quad \text{GCF of 2}$$

For first factors, begin with $2x$ and $2x$ (closest factors). For last factors, begin with -5 and -2 (closest factors and the product is positive; and since the middle term is negative, both factors need to be negative).

$$(2x - 5)(2x - 2)$$

Shortcut 3 eliminates this possibility.

Now, try -1 and -10 for last factors.

$$(2x - 1)(2x - 10)$$

Shortcut 3 eliminates this possibility.

Now, try $1x$ and $4x$ for first factors and go back to -5 and -2 as last factors.

$$(x - 5)(4x - 2)$$

Shortcut 3 eliminates this possibility. But because x and $4x$ are different factors, switching the -5 and -2 produces different results, as shown in the following:

$$\text{outer} = -5x$$

$$(x - 2)(4x - 5) \quad \text{sum of outer and inner} = -13x$$

$$\text{inner} = -8x$$

Therefore, $\qquad 8x^2 - 26x + 20 = 2(x - 2)(4x - 5)$

Square Trinomials

Recall that

$$\underbrace{\left(x+y\right)^2}_{\substack{\text{the square} \\ \text{of a} \\ \text{binomial}}} = \underbrace{\left(x\right)^2 + 2\left(x\right)\left(y\right) + \left(y\right)^2}_{\text{square trinomial}}$$

Therefore, if a trinomial is of the form $(x)^2 + 2(x)(y) + (y)^2$, it can be factored into the square of a binomial.

Example 18: Is $4x^2 - 20x + 25$ a square trinomial? If so, factor it into the square of some binomial.

$$4x^2 = (2x)^2 \quad \text{and} \quad 25 = (-5)^2 \quad \text{and} \quad -20x = 2(2x)(-5)$$

So it is a square trinomial, which factors as follows.

$$4x^2 - 20x + 25 = (2x - 5)$$

Example 19: Is $x^2 + 10x + 9$ a square trinomial?

$$x^2 = (x)^2 \quad \text{and} \quad 9 = 3^2 \quad \text{but} \quad 10x \neq 2(x)(3)$$

So it is not a square trinomial. But $x^2 + 10x + 9$ is factorable.

$$x^2 + 10x + 9 = (x + 1)(x + 9)$$

Factoring by Regrouping

To attempt to factor a polynomial of four or more terms with no common factor, first rewrite it in groups. Each group may then be separately factored, and the resulting expression may lend itself to further factorization.

Example 20: Factor $ay + az + by + bz$.

This polynomial has four terms with no common factor. It could be put into either two groups of two terms or two groups with three terms in one group and one term in the other group. One such arrangement is

$$(ay + az) + (by + bz)$$

$$ay + az = a(y + z) \quad \text{GCF of } a$$

$$by + bz = b(y + z) \quad \text{GCF of } b$$

So

$$(ay + az) + (by + bz) = a(\underline{y + z}) + b(\underline{y + z}) \qquad \text{GCF of } (y + z)$$
$$= (y + z)(a + b)$$

Example 21: Factor $x^2 + 2xy + y^2 - z^2$.

This polynomial has four terms with no common factor. Also notice that

$$x^2 + 2xy + y^2 = (x + y)^2 \qquad \text{square trinomial}$$

$$\begin{aligned}
x^2 + 2xy + y^2 - z^2 &= (x^2 + 2xy + y^2) - z^2 \\
&= (x + y)^2 - z^2 \quad \text{difference of squares} \\
&= [(x + y) + z][(x + y) - z] \\
&= (x + y + z)(x + y - z)
\end{aligned}$$

Example 22: Factor $x^2 - y^2 + x^3 + y^3$

$$x^2 - y^2 + x^3 + y^3 = \underbrace{\left(x^2 - y^2\right)}_{\substack{\text{difference} \\ \text{of squares}}} + \underbrace{\left(x^3 + y^3\right)}_{\substack{\text{sum of} \\ \text{cubes}}}$$

$$= [(\underline{x + y})(x - y)] + [(x + y)(x^2 - xy + y^2)]$$

$\text{GCF} = (x + y)$
$$= (x + y)[(x - y) + (x^2 - xy + y^2)]$$
$$= (x + y)(x^2 + x - xy - y + y^2)$$

Summary of Factoring Techniques

■ For all polynomials, first factor out the greatest common factor (GCF).

■ For a binomial, check to see if it is any of the following:

    (a) difference of squares: $x^2 - y^2 = (x + y)(x - y)$

    (b) difference of cubes: $x^3 - y^3 = (x - y)(x^2 + xy + y^2)$

    (c) sum of cubes: $x^3 + y^3 = (x + y)(x^2 - xy + y^2)$

■ For a trinomial, check to see if it is either of the following forms:

(a) $x^2 + bx + c$:

If so, find two integers whose product is c and whose sum is b. For example,

$$x^2 + 8x + 12 = (x + 2)(x + 6)$$

since $(2)(6) = 12$ and $2 + 6 = 8$

(b) $ax^2 + bx + c$:

If so, find two binomials so that

the product of first terms $= ax^2$

the product of last terms $= c$

the sum of outer and inner products $= bx$

See the following where the product of first terms $= (3x)(2x) = 6x^2$, the product of last terms $= (2)(-5) = -10$, and the sum of outer and inner products $= (3x)(-5) + 2(2x) = -11x$.

$$6x^2 - 11x - 10 - (3x + 2)(2x - 5)$$

(c) $x^2 + 2xy + y^2 = (x + y)^2$

 $x^2 - 2xy + y^2 = (x - y)^2$ square trinomials

■ For polynomials with four or more terms, regroup, factor each group, and then find a pattern as in Steps 1 through 3.

Solving Equations by Factoring

Factoring is a method that can be used to solve equations of degree higher than 1. This method uses the zero product rule.

> *Zero Product Rule*
>
> If $(a)(b) = 0$, then
>
> Either $(a) = 0$, $(b) = 0$, or both.

Example 23: Solve $x(x + 3) = 0$.

$$x(x + 3) = 0$$

Apply the zero product rule.

$x = 0$	or	$x + 3 = 0$
$x = 0$	or	$x = -3$

Check the solution.

$x(x + 3) = 0$	$x(x + 3) = 0$
$0(0 + 3) \overset{?}{=} 0$	$-3(-3 + 3) \overset{?}{=} 0$
$0 = 0$ ✓	$0 = 0$ ✓

The solution is $x = 0$ or $x = -3$.

Example 24: Solve $x^2 - 5x + 6 = 0$.

$$x^2 - 5x + 6 = 0$$

Factor.

$$(x - 2)(x - 3) = 0$$

Apply the zero product rule.

$x - 2 = 0$	or	$x - 3 = 0$
$x = 2$		$x = 3$

The check is left to you. The solution is $x = 2$ or $x = 3$.

Example 25: Solve $3x(2x - 5) = -4(4x - 3)$.

$$3x(2x - 5) = -4(4x - 3)$$

Distribute.

$$6x^2 - 15x = -16x + 12$$

Get all terms on one side, leaving zero on the other, in order to apply the zero product rule.

$$6x^2 + x - 12 = 0$$

Factor.

$$(3x - 4)(2x + 3) = 0$$

Apply the zero product rule.

$$3x - 4 = 0 \qquad \text{or} \qquad 2x + 3 = 0$$
$$3x = 4 \qquad\qquad\qquad 2x = -3$$
$$x = \tfrac{4}{3} \qquad\qquad\qquad x = -\tfrac{3}{2}$$

The check is left to you. The solution is $x = \tfrac{4}{3}$ or $x = -\tfrac{3}{2}$.

Example 26: Solve $2y^3 = 162y$.

$$2y^3 = 162y$$

Get all terms on one side of the equation.

$$2y^3 - 162y = 0$$

Factor (GCF).

$$2y(y^2 - 81) = 0$$

Continue to factor (difference of squares).

$$2y(y + 9)(y - 9) = 0$$

Apply the zero product rule.

$$2y = 0 \quad \text{or} \quad y + 9 = 0 \qquad \text{or} \qquad y - 9 = 0$$
$$y = 0 \qquad\qquad y = -9 \qquad\qquad\qquad y = 9$$

The check is left to you. The solution is $y = 0$ or $y = -9$ or $y = 9$.

Chapter Checkout

1. Factor completely:

   a) $36x^2 - 25$

   b) $20x^2 - 80$

2. Factor:

   a) $x^2 + 8x + 15$

   b) $x^2 + 2x - 8$

   c) $x^2 - 5x + 6$

3. Solve for x:

$$x^2 + 2x - 24 = 0$$

Answers: 1. a. $(6x+5)(6x-5)$ b. $20(x+2)(x-2)$ 2. a. $(x+3)(x+5)$
b. $(x+4)(x-2)$ c. $(x-2)(x-3)$ 3. $x=4, x=-6$

Chapter 7

RATIONAL EXPRESSIONS

Chapter Check-In

- ❑ Simplifying rational expressions
- ❑ Doing arithmetic by using rational expressions
- ❑ Working with complex fractions
- ❑ Solving and graphing rational functions
- ❑ Working with proportions and variations

Rational expression is a fancy way of saying fraction. Of course, a fraction may also be perceived as being a division example, wherein the numerator is being divided by the denominator. While the basic rules of arithmetic of fractions apply to the rational expressions treated within this chapter, having polynomials in the numerators and/or denominators does, at times, require some fancy footwork. This is even more true when there is a fraction in the numerator and another one in the denominator (a complex fraction). The chapter also examines the concepts of direct, inverse, and joint variation, and deals with solving rational equations and graphing rational functions.

Examples of Rational Expressions

The quotient of two polynomials is called a rational expression. The denominator of a rational expression can never have a zero value. The following are examples of rational expressions:

$$\frac{5}{6}, \ \frac{3x+8}{2x-9}, \ \frac{x^2-7x+3}{12}, \ \frac{9x}{x^2+5x+6}, \ 6x+5$$

The last example, $6x + 5$, could be expressed as

$$\frac{6x+5}{1}$$

Therefore, it satisfies the definition of a rational expression.

Simplifying Rational Expressions

To simplify a rational expression:

1. **Completely factor numerators and denominators.**
2. **Reduce common factors.**

Example 1: Simplify $\dfrac{5x+15}{4x+12}$.

$$\frac{5x+15}{4x+12} = \frac{5\,(\overset{1}{\cancel{x+3}})}{4\,(\underset{1}{\cancel{x+3}})} = \frac{5}{4}$$

Example 2: Simplify $\dfrac{x^2-16}{x^3+64}$.

$$\frac{x^2-16}{x^3+64} = \frac{4\,(\overset{1}{\cancel{x+4}})\,(x-4)}{4\,(\underset{1}{\cancel{x+4}})\left(x^2-4x+16\right)} = \frac{x-4}{x^2-4x+16}$$

Whenever possible, try to write all polynomials in descending order with a positive leading coefficient. To have a positive leading coefficient, occasionally −1 has to be factored out of the polynomial.

Example 3: Simplify $\dfrac{4x^2-25}{5-2x}$.

$$\frac{4x^2-25}{5-2x} = \frac{(2x+5)(\overset{1}{\cancel{2x-5}})}{-1\,(\underset{1}{\cancel{2x-5}})} = \frac{2x+5}{-1} = -2x-5$$

Example 4: Simplify $\dfrac{x^2-7x+12}{15-2x-x^2}$.

$$\frac{x^2-7x+12}{15-2x-x^2} = \frac{(x-3)(x-4)}{-1\left(x^2+2x-15\right)} = \frac{(\overset{1}{\cancel{x-3}})\,(x-4)}{-1\,(\underset{1}{\cancel{x-3}})\,(x+5)} = \frac{x-4}{-1(x+5)}$$

This answer can be expressed in other ways.

$$\frac{x-4}{-1(x+5)} = \frac{x-4}{-x-5}, \quad \frac{x-4}{-1(x+5)} \cdot \frac{-1}{-1} = \frac{-x+4}{x+5}, \text{ or } -\frac{x-4}{x+5}$$

Multiplying Rational Expressions

To multiply rational expressions:

1. **Completely factor all numerators and denominators.**
2. **Reduce all common factors.**
3. **Either multiply the denominators and numerators together or leave the answer in factored form.**

Example 5: Simplify $\dfrac{x^2 - 8x + 12}{x^2 - 16} \cdot \dfrac{4x + 16}{x^2 - 4x + 4}$.

$$\frac{x^2 - 8x + 12}{x^2 - 16} \cdot \frac{4x + 16}{x^2 - 4x + 4}$$

$$= \frac{(x - 2)\ (x - 6)}{(x + 4)\ (x - 4)} \cdot \frac{4\ (x + 4)}{(x - 2)^2}$$

$$= \frac{4\,(x - 6)}{(x - 4)(x - 2)}$$

This last answer could be either left in its factored form or multiplied out. If multiplied out, it becomes

$$\frac{4x - 24}{x^2 - 6x + 8}$$

Example 6: Simplify $(x^2 - 2x) \cdot \dfrac{x}{x^2 - 5x + 6}$.

$$\left(x^2 - 2x\right) \cdot \frac{x}{x^2 - 5x + 6}$$

$$= \frac{x\,(x - 2)}{1} \cdot \frac{x}{(x - 2)\ (x - 3)}$$

$$= \frac{x^2}{x - 3}$$

Example 7: Simplify $\dfrac{9 - x^2}{x^2 + 6x + 9} \cdot \dfrac{3x + 9}{3x - 9}$.

$$\frac{9-x^2}{x^2+6x+9} \cdot \frac{3x+9}{3x-9}$$

$$= \frac{-1\,(x+3)\,(x-3)}{(x+3)^2} \cdot \frac{3\,(x+3)}{3\,(x-3)}$$

$$= \frac{-1}{1} = -1$$

Dividing Rational Expressions

From arithmetic, division of fractions involves multiplying by the reciprocal of the divisor.

Example 8: Simplify $\dfrac{x^2-9x-10}{x^2+x-6} \div \dfrac{x^2-1}{x^2-4}$.

$$\frac{x^2-9x-10}{x^2+x-6} \div \frac{x^2-1}{x^2-4} = \frac{x^2-9x-10}{x^2+x-6} \cdot \frac{x^2-4}{x^2-1}$$

Multiply by the reciprocal of $\dfrac{x^2-1}{x^2-4}$

$$\frac{x^2-9x-10}{x^2+x-6} \cdot \frac{x^2-4}{x^2-1}$$

$$= \frac{(x-10)(x+1)}{(x-2)\,(x+3)} \cdot \frac{(x+2)(x-2)}{(x+1)\,(x-1)}$$

$$= \frac{(x-10)(x+2)}{(x+3)(x-1)}$$

If this last answer is multiplied out, it is

$$\frac{x^2-8x-20}{x^2+2x-3}$$

Example 9: Simplify $\dfrac{2x^2-2x-4}{x^2+2x-8} \div \dfrac{4x^2-100}{x^2-x-20} \cdot \dfrac{3x^2+15x}{x+1}$.

Since only the middle expression is doing the dividing, it is the only one whose reciprocal is used.

$$\frac{2x^2 - 2x - 4}{x^2 + 2x - 8} \div \frac{4x^2 - 100}{x^2 - x - 20} \cdot \frac{3x^2 + 15x}{x + 1}$$

$$= \frac{2x^2 - 2x - 4}{x^2 + 2x - 8} \cdot \frac{x^2 - x - 20}{4x^2 - 100} \cdot \frac{3x^2 + 15x}{x + 1}$$

$$= \frac{\overset{1}{\cancel{2}}(x - 2)(x + 1)}{(x + 4)(x - 2)} \cdot \frac{(x + 4)(x - 5)}{\underset{2}{\cancel{4}}(x + 5)(x - 5)} \cdot \frac{3x(x + 5)}{x + 1}$$

$$= \frac{3x}{2}$$

Adding and Subtracting Rational Expressions

To add or subtract rational expressions with the same denominators:

1. **Add or subtract the numerators as indicates.**
2. **Keep the common denominator.**
3. **Simplify the resulting rational expression if possible.**

Example 10: Simplify $\dfrac{4}{5x} + \dfrac{1}{5x}$.

$$\frac{4}{5x} + \frac{1}{5x} = \frac{4 + 1}{5x} = \frac{\overset{1}{\cancel{5}}}{\underset{1}{\cancel{5}}x} = \frac{1}{x}$$

Example 11: Simplify $\dfrac{x^2 + 5x + 1}{x + 3} - \dfrac{4x - 5}{x + 3} + \dfrac{7x + 9}{x + 3}$.

$$\frac{x^2 + 5x + 1}{x + 3} - \frac{4x - 5}{x + 3} + \frac{7x + 9}{x + 3}$$

$$= \frac{x^2 + 5x + 1 - (4x - 5) + 7x + 9}{x + 3}$$

$$= \frac{x^2 + 5x + 1 - 4x + 5 + 7x + 9}{x + 3}$$

$$= \frac{x^2 + 8x + 15}{x + 3}$$

$$= \frac{(x + 3)(x + 5)}{x + 3}$$

$$= x + 5$$

To add or subtract rational expressions with *different* denominators:

1. **Completely factor each denominator.**
2. **Find the least common denominator (LCD) for all the denominators by multiplying together the different prime factors with the greatest exponent for each factor.**
3. **Rewrite each fraction so it has the LCD as its denominator by multiplying each fraction by the value 1 in an appropriate form.**
4. **Combine numerators as indicated and keep the LCD as the denominator.**
5. **Simplify the resulting rational expression if possible.**

Example 12: Simplify $\frac{5}{x} + \frac{2}{y}$.

Completely factor each denominator.

x and y are already prime factors.

Find the least common denominator (LCD) for all the denominators.

The LCD = xy.

Rewrite each fraction so it has the LCD as its denominator.

$$\frac{5}{x} + \frac{2}{y}$$

$$= \frac{5}{x} \cdot \left(\frac{y}{y}\right) + \frac{2}{y} \cdot \left(\frac{x}{x}\right)$$

$$= \frac{5y}{xy} + \frac{2x}{xy}$$

Combine numerators and keep the LCD as the denominator.

$$= \frac{5y + 2x}{xy}$$

This rational expression cannot be simplified further. Therefore,

$$\frac{5}{x} + \frac{2}{y} = \frac{5y + 2x}{xy}$$

Example 13: Simplify $\dfrac{4}{x^2 - 16} + \dfrac{3}{x^2 + 8x + 16}$.

Factor each denominator.

$$x^2 - 16 = (x + 4)(x - 4)$$
$$x^2 + 8x + 16 = (x + 4)^2$$
$$\frac{4}{x^2 - 16} + \frac{3}{x^2 + 8x + 16} = \frac{4}{(x+4)(x-4)} + \frac{3}{(x+4)^2}$$

Find the LCD.

$$\text{The LCD} = (x - 4)(x + 4)^2.$$

Rewrite fraction so that LCD is denominator.

$$\frac{4}{x^2 - 16} + \frac{3}{x^2 + 8x + 16}$$

$$= \frac{4}{(x+4)(x-4)} + \frac{3}{(x+4)^2}$$

$$= \frac{4}{(x+4)(x-4)} \cdot \left[\frac{(x+4)}{(x+4)}\right] + \frac{3}{(x+4)^2} \cdot \left[\frac{(x-4)}{(x-4)}\right]$$

$$= \frac{4x + 16}{(x-4)(x+4)^2} + \frac{3x - 12}{(x-4)(x+4)^2}$$

Combine numerators and keep LCD as denominator.

$$= \frac{7x + 4}{(x-4)(x+4)^2}$$

This rational expression cannot be simplified further. Therefore,

$$\frac{4}{x^2 - 16} + \frac{3}{x^2 + 8x + 16} = \frac{7x + 4}{(x-4)(x+4)^2}$$

Example 14: Simplify $\frac{8x}{x-3} + \frac{5}{9 - x^2}$.

$(x - 3)$ is a prime factor.
Rewrite in descending order.

$$9 - x^2 = -x^2 + 9$$

Factor out −1 so the leading coefficient is positive.

$$= -1(x^2 - 9)$$

$$= -1(x - 3)(x + 3)$$

The LCD = $(x - 3)(x + 3)$. [The LCD could also have been
−1$(x - 3)(x + 3)$.]

$$\frac{8x}{x-3} + \frac{5}{9-x^2}$$

$$= \frac{8x}{(x-3)} + \frac{5}{-1(x-3)(x+3)}$$

$$= \frac{8x}{(x-3)} \cdot \left[\frac{(x+3)}{(x+3)}\right] + \frac{5}{-1(x-3)(x+3)} \cdot \left(\frac{-1}{-1}\right)$$

$$= \frac{8x^2 + 24x}{(x-3)(x+3)} + \frac{-5}{(x-3)(x+3)}$$

$$= \frac{8x^2 + 24x - 5}{(x-3)(x+3)}$$

This rational expression cannot be simplified further. Therefore,

$$\frac{8x}{x-3} + \frac{5}{9-x^2} = \frac{8x^2 + 24x - 5}{(x-3)(x+3)}$$

Example 15: Simplify $\dfrac{2x}{x^2-4} - \dfrac{1}{x^2-3x+2} + \dfrac{x+1}{x^2+x-2}$.

Factor each denominator.

$$x^2 - 4 = (x+2)(x-2)$$
$$x^2 - 3x + 2 = (x-2)(x-1)$$
$$x^2 + x - 2 = (x+2)(x-1)$$

The LCD $= (x+2)(x-2)(x-1)$.

Rewrite the fraction so the LCD is the denominator.

$$\frac{2x}{x^2-4} - \frac{1}{x^2-3x+2} + \frac{x+1}{x^2+x-2}$$

$$= \frac{2x}{(x+2)(x-2)} - \frac{1}{(x-2)(x-1)} + \frac{x+1}{(x+2)(x-1)}$$

$$= \frac{2x}{(x+2)(x-2)} \cdot \left[\frac{(x-1)}{(x-1)}\right] - \frac{1}{(x-2)(x-1)} \cdot \left[\frac{(x+2)}{(x+2)}\right] +$$

$$\frac{(x+1)}{(x+2)(x-1)} \cdot \left[\frac{(x-2)}{(x-2)}\right]$$

$$= \frac{2x^2 - 2x}{(x+2)(x-2)(x-1)} - \frac{x+2}{(x+2)(x-2)(x-1)} +$$

$$\frac{x^2 - 2x + x - 2}{(x+2)(x-2)(x-1)}$$

$$= \frac{2x^2 - 2x - x - 2 + x^2 - 2x + x - 2}{(x+2)(x-2)(x-1)}$$

$$= \frac{3x^2 - 4x - 4}{(x+2)(x-2)(x-1)}$$

This rational expression can be simplified.

$$= \frac{(3x+2)(x - 2)^{1}}{(x+2)(x - 2)_{1}\,(x-1)}$$

$$= \frac{(3x+2)}{(x+2)(x-1)}$$

Therefore,

$$\frac{2x}{x^2-4} - \frac{1}{x^2-3x+2} + \frac{x+1}{x^2+x-2} = \frac{3x+2}{(x+2)(x-1)}$$

Complex Fractions

If the numerator or denominator or both contain fractions, then the expression is called a **complex fraction.** The fraction bar indicates division. To simplify a complex fraction:

1. **Simplify both numerator and denominator expressions into single fraction expressions.**
2. **Divide the numerator by the denominator and simplify if possible.**

Example 16: Simplify $\dfrac{x + 1 - \dfrac{6}{x}}{\dfrac{1}{x}}$.

Simplify numerator and denominator.

$$x + 1 - \frac{6}{x} = \frac{x^2 + x - 6}{x}$$

$\frac{1}{x}$ is already a single fraction expression.

$$\frac{x + 1 - \dfrac{6}{x}}{\dfrac{1}{x}} = \frac{\dfrac{x^2 + x - 6}{x}}{\dfrac{1}{x}}$$

Divide numerator by denominator and simplify.

$$= \frac{x^2 + x - 6}{x} \div \frac{1}{x}$$

$$= \frac{x^2 + x - 6}{x} \cdot \frac{x}{1}$$

$$= \frac{(x + 3)(x - 2)}{\cancel{x}_1} \cdot \frac{\cancel{x}^1}{1}$$

$$= (x + 3)(x - 2) \text{ or } x^2 + x - 6$$

Therefore,

$$\frac{x + 1 - \dfrac{6}{x}}{\dfrac{1}{x}} = (x + 3)(x - 2)$$

Example 17: Simplify $\dfrac{\dfrac{1}{x^2} - \dfrac{1}{y^2}}{\dfrac{1}{x^3} + \dfrac{1}{y^3}}$.

Simplify numerator and denominator.

$$\frac{1}{x^2} - \frac{1}{y^2} = \frac{y^2 - x^2}{x^2 y^2}$$

$$\frac{1}{x^3} + \frac{1}{y^3} = \frac{y^3 + x^3}{x^3 y^3}$$

$$\frac{\dfrac{1}{x^2} - \dfrac{1}{y^2}}{\dfrac{1}{x^3} + \dfrac{1}{y^3}} = \frac{\dfrac{y^2 - x^2}{x^2 y^2}}{\dfrac{y^3 + x^3}{x^3 y^3}}$$

Divide numerator by denominator and simplify.

$$= \frac{y^2 - x^2}{x^2 y^2} \div \frac{y^3 + x^3}{x^3 y^3}$$

$$= \frac{\overbrace{y^2 - x^2}^{\substack{\text{difference} \\ \text{of squares}}}}{x^2 y^2} \cdot \frac{x^3 y^3}{\underbrace{y^3 + x^3}_{\substack{\text{sum of} \\ \text{cubes}}}}$$

$$= \frac{(\overset{1}{\cancel{y + x}})(y - x)}{\cancel{x^2 y^2}} \cdot \frac{\cancel{x^3 y^3}^{xy}}{(\underset{1}{\cancel{y + x}})(y^2 - xy + x^2)}$$

$$= \frac{xy(y - x)}{y^2 - xy + x^2}$$

Therefore,

$$\frac{\dfrac{1}{x^2} - \dfrac{1}{y^2}}{\dfrac{1}{x^3} + \dfrac{1}{y^3}} = \frac{xy(y - x)}{y^2 - xy + x^2}$$

Solving Rational Equations

An equation involving rational expressions is called a **rational equation.** To solve such an equation:

1. **Completely factor all denominators.**
2. **Multiply both sides of the equation by the least common denominator (LCD).**
3. **Solve the resulting equation.**
4. **Eliminate any solutions that would make the LCD become zero. These solutions are called extraneous solutions.**
5. **Check the remaining solutions.**

Example 18: Solve $\dfrac{5}{x} + \dfrac{4}{x + 3} = \dfrac{8}{x^2 + 3x}$.

Factor all denominators.

$$\frac{5}{x} + \frac{4}{x + 3} = \frac{8}{x(x + 3)} \qquad [\text{LCD} = x(x + 3)]$$

Multiply both sides by LCD.

$$x(x+3)\left[\frac{5}{x}+\frac{4}{x+3}\right]=x(x+3)\left[\frac{8}{x(x+3)}\right]$$

$$\frac{\cancel{x}(x+3)}{1}\cdot\frac{5}{\cancel{x}}+\frac{x(\cancel{x+3})}{1}\cdot\frac{4}{\cancel{x+3}}=\frac{\cancel{x}(\cancel{x+3})}{1}\cdot\frac{8}{\cancel{x}(\cancel{x+3})}$$

Solve equation.

$$5x+15+4x=8$$
$$9x=-7$$
$$x=-\tfrac{7}{9}$$

Since $-\tfrac{7}{9}$ is not an extraneous solution because it does not make the LCD zero, all that is needed is to verify that $-\tfrac{7}{9}$ is a solution to the original equation.

The check is left to you.

Example 19: Solve $\frac{x+1}{5}-2=\frac{-4}{x}$.

Factor denominators.

$$\frac{x+1}{5}-2=\frac{-4}{x}\qquad(\text{LCD}=5x)$$

Multiply both sides by LCD.

$$5x\cdot\left(\frac{x+1}{5}-2\right)=5x\cdot\left(\frac{-4}{x}\right)$$

$$\frac{\cancel{5}x}{1}\cdot\left(\frac{x+1}{\cancel{5}}\right)-\frac{5x}{1}\cdot(2)=\frac{5\cancel{x}}{1}\cdot\left(\frac{-4}{\cancel{x}}\right)$$

Solve equation.

$$x^2+x-10x=-20$$

Get all terms on one side and solve by factoring.

$$x^2 - 9x + 20 = 0$$

$$(x - 4)(x - 5) = 0$$

$$x = 4 \quad \text{or} \quad x = 5$$

Since neither answer is an extraneous solution, each must be checked with the original equation. The check of these solutions is left to you.

Proportion, Direct Variation, Inverse Variation, Joint Variation

This section defines what proportion, direct variation, inverse variation, and joint variation are and explains how to solve such equations.

Proportion

A **proportion** is an equation stating that two rational expressions are equal. Simple proportions can be solved by applying the cross products rule.

> *Cross Products Rule*
>
> If $\frac{a}{b} = \frac{c}{d}$, then $ad = bc$.

More involved proportions are solved as rational equations.

Example 20: Solve $\frac{x}{15} = \frac{7}{8}$.

$$\frac{x}{15} = \frac{7}{8}$$

Apply the cross products rule.

$$8x = (15)(7)$$

$$8x = 105$$

$$x = \frac{105}{8}$$

The check is left to you.

Example 21: Solve $\frac{x+5}{3} = \frac{7x-2}{5}$.

$$\frac{x+5}{3} = \frac{7x-2}{5}$$

Apply the cross products rule.

$$5(x + 5) = 3(7x - 2)$$
$$5x + 25 = 21x - 6$$
$$-16x = -31$$
$$x = \tfrac{31}{16}$$

The check is left to you.

Example 22: Solve $\dfrac{41x - 12}{x^2 - 16} = \dfrac{4x + 3}{x - 4}$.

$$\frac{41x - 12}{x^2 - 16} = \frac{4x + 3}{x - 4} \qquad [\text{LCD} = (x + 4)(x - 4)]$$

$$(x + 4)(x - 4) \left[\frac{41x - 12}{(x + 4)\,(x - 4)} \right] = (x + 4)(x - 4) \left[\frac{4x + 3}{x - 4} \right]$$

$$41x - 12 = 4x^2 + 19x + 12$$
$$0 = 4x^2 - 22x + 24$$
$$0 = 2\left(2x^2 - 11x + 12\right)$$
$$0 = 2\left(2x - 3\right)\left(x - 4\right)$$
$$2x - 3 = 0 \qquad \text{or} \quad x - 4 = 0$$
$$x = \tfrac{3}{2} \qquad\qquad x = 4$$

However, $x = 4$ is an extraneous solution, because it makes the LCD become zero. To check to see if $x = \tfrac{3}{2}$ is a solution is left to you.

Direct variation

The phrase "y **varies directly** as x" or "y is directly proportional to x" can be translated in two ways.

- $\dfrac{y}{x} = k$ for some constant k.

  The k is called the **constant of proportionality.** This translation is used when the constant is the desired result.

- $\dfrac{y_1}{x_1} = \dfrac{y_2}{x_2}$

  This translation is used when the desired result is either an original or new value of x or y.

Example 23: If y varies directly as x, and y = 10 when x = 7, find the constant of proportionality.

Use $\dfrac{y}{x} = k$

$\dfrac{10}{7} = k$

The constant of proportionality is $\frac{10}{7}$.

Example 24: If y varies directly as x, and y = 10 when x = 7, find y when x = 12.

Use $\dfrac{y_1}{x_1} = \dfrac{y_2}{x_2}$

$\dfrac{10}{7} = \dfrac{y}{12}$

Apply the cross products rule.

$$7y = 120$$

$$y = \tfrac{120}{7}$$

Inverse variation

The phrase "y **varies inversely** as x" or "y is inversely proportional to x" is translated in two ways.

■ $yx = k$ for some constant k, called the constant of proportionality. Use this translation if the constant is desired.

■ $y_1x_1 = y_2x_2$.

Use this translation if a value of x or y is desired.

Example 25: If y varies inversely as x, and y = 4 when x = 3, find the constant of proportionality.

Use $$yx = k$$
$$(4)(3) = k$$
$$k = 12$$

The constant is 12.

Example 26: If y varies inversely as x, and y = 9 when x = 2, find y when x = 3.

Use $$y_1x_1 = y_2x_2$$
$$(9)(2) = (y)(3)$$
$$3y = 18$$
$$y = 6$$

Joint variation

If one variable varies as the product of other variables, it is called **joint variation.** The phrase "*y* **varies jointly** as *x* and *z*" is translated in two ways.

- $\dfrac{y}{xz} = k$ if the constant is desired
- $\dfrac{y_1}{x_1 z_1} = \dfrac{y_2}{x_2 z_2}$ if one of the variables is desired

Example 27: If *y* varies jointly as *x* and *z*, and *y* = 10 when *x* = 4 and *x* = 5, find the constant of proportionality.

Use
$$\frac{y}{xz} = k$$
$$\frac{10}{(4)(5)} = k$$
$$\tfrac{10}{20} = k$$
$$k = \tfrac{1}{2}$$

Example 28: If *y* varies jointly as *x* and *z*, and *y* = 12 when *x* = 2 and *z* = 3, find *y* when *x* = 7 and *z* = 4.

$$\frac{y_1}{x_1 z_1} = \frac{y_2}{x_2 z_2}$$
$$\frac{12}{(2)(3)} = \frac{y}{(7)(4)}$$
$$\frac{12}{6} = \frac{y}{28}$$
$$6y = 336$$
$$y = 56$$

Occasionally, a problem involves both direct and inverse variations. Suppose that *y* varies directly as *x* and inversely as *z*. This involves three variables and can be translated in two ways:

- $\dfrac{yz}{x} = k$ if the constant is desired
- $\dfrac{y_1 z_1}{x_1} = \dfrac{y_2 z_2}{x_2}$

Example 29: If *y* varies directly as *x* and inversely as *z*, and *y* = 5 when *x* = 2 and *z* = 4, find *y* when *x* = 3 and *z* = 6.

$$\frac{y_1 z_1}{x_1} = \frac{y_2 z_2}{x_2}$$

$$\frac{(5)(4)}{2} = \frac{(y)(6)}{3}$$

$$12y = 60$$

$$y = 5$$

Graphing Rational Functions

If $f(x)$ represents a rational expression, then $y = f(x)$ is a **rational function.** To graph a rational function, first find values for which the function is undefined. A function is undefined for any values that would make any denominator become zero. Dashed lines are drawn on the graph for any values for which the rational function is undefined. These lines are called **asymptote lines.** The graph of the rational function will get close to these asymptote lines but will never intersect them.

Example 30: Graph $y = \frac{2}{x+1}$.

First, the function is undefined when $x = -1$. So the graph of $x = -1$ becomes a vertical asymptote.

Second, find if any horizontal asymptotes exist. To do this, solve the equation for x and see if there are any values for y that would make the new equation undefined. It may not always be easy to solve a rational function for x. For those cases, other methods are used involving the idea of limits, which is discussed in calculus.

$$y = \frac{2}{x+1}$$

Apply the cross products rule.

$$y(x + 1) = 2$$

$$xy + y = 2$$

$$xy = 2 - y$$

$$x = \frac{2-y}{y}$$

This equation is undefined when $y = 0$. The graph of $y = 0$ becomes a horizontal asymptote. The graph of $x = -1$ and $y = 0$ is shown in Figure 7-1.

Figure 7-1 $x = -1$ is an asymptote.

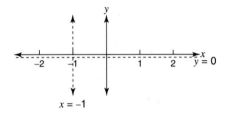

Third, plot the following points on each side of each asymptote line and use them to sketch the graph (see Figure 7-2).

x	$y = \dfrac{2}{x+1}$	y
-4	$y = 2/(-4 + 1)$	$-\frac{2}{3}$
-3	$y = 2/(-3 + 1)$	-1
-2	$y = 2/(-2 + 1)$	-2
-1	$y = 2/(-1 + 1)$	undefined
0	$y = 2/(0 + 1)$	2
1	$y = 2/(1 + 1)$	1
2	$y = 2/(2 + 1)$	$\frac{2}{3}$
3	$y = 2/(3 + 1)$	$\frac{1}{2}$

Figure 7-2   Graphs don't hit asymptotes.

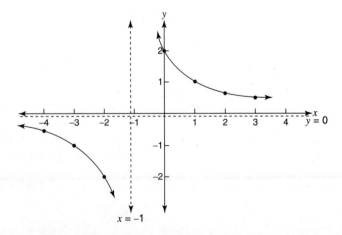

Example 31: Graph $y = \dfrac{x}{x+2}$.

First, x = −2 is the vertical asymptote.

Second, find any horizontal asymptotes that exist by solving the equation for x.

$$y = \frac{x}{x+2}$$

Apply the cross products rule.

$$y(x+2) = x$$
$$xy + 2y = x$$
$$xy - x = -2y$$
$$x(y - 1) = -2y$$
$$x = \frac{-2y}{y - 1}$$

This equation is undefined when $y = 1$, so $y = 1$ is a horizontal asymptote as shown in Figure 7-3.

Figure 7-3 $y = 1$ is a horizontal asymptote.

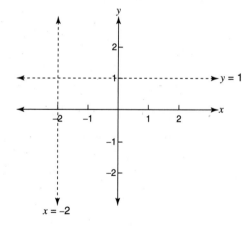

Third, plot these points on either side of each asymptote line and sketch the graph as shown in Figure 7-4.

x	$y = \dfrac{x}{x+2}$	y
-5	$y = -5/(-5+2)$	$\frac{5}{3}$
-4	$y = -4/(-4+2)$	2
-3	$y = -3/(-3+2)$	3
-2	$y = -2/(-2+2)$	undefined
-1	$y = -1/(-1+2)$	-1
0	$y = 0/(0+2)$	0
1	$y = 1/(1+2)$	$\frac{1}{3}$
2	$y = 2/(2+2)$	$\frac{1}{2}$
3	$y = 3/(3+2)$	$\frac{3}{5}$

Figure 7-4   The expression is plotted.

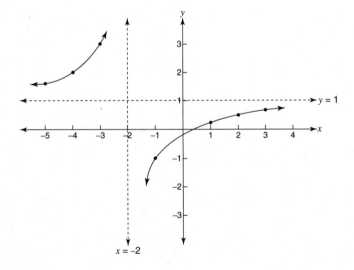

Chapter Checkout

1. Simplify the following:

$$\frac{9y^2 - 36}{6 - 3y}$$

2. Simplify the following::

$$\frac{9}{x^2 - 9} + \frac{5}{x^2 + 6x + 9}$$

3. Solve the following::

$$\frac{x + 7}{3} = \frac{3x - 2}{5}$$

4. If y varies inversely as x, and $y = 6$ when $x = 3$, find the constant of proportionality.

Answers: 1. $-3y - 6$ or $-3(y + 2)$ 2. $\dfrac{2(7x + 6)}{(x + 3)^2(x - 3)}$

3. $x = \dfrac{41}{4}$ 4. $k = 18$

Chapter 8

RELATIONS AND FUNCTIONS

Chapter Check-In

❑ Defining relation, range, domain, and function

❑ Exploring function notation and composition

❑ Learning the algebra of functions

❑ Creating inverse functions

The horizontal coordinates of a set of ordered pairs constitutes the domain of a relation, while the range is determined by the vertical coordinates. A relation in which none of the domain numbers appear more than once is called a "function." If two functions have a common domain, they can be acted upon arithmetically. In addition to the foregoing, this chapter also deals with the fact that the inverse of a function may or may not be a function.

Basic Definitions

Following are definitions with which you should be familiar as you work with relations and functions.

Relation

A **relation** is a set of ordered pairs that can be represented by a diagram, graph, or sentence. Figure 8-1 shows several examples of relations.

Figure 8-1 Examples of relations.

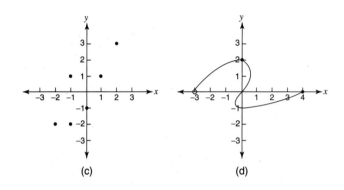

{(4,1),(3,2),(−1,6)}

(a)

(b)

(c)

(d)

(e) $y = 2x + 3$

Domain and range

The set of all first numbers of the ordered pairs in a relation is called the **domain** of the relation. The set of all second numbers of the ordered pairs in a relation is called the **range** of the relation. The values in the domain and range are usually listed from least to greatest.

Example 1: Find the domain and range of item (a) in Figure 8-1.

$$\text{domain} = \{-1, 3, 4\}$$
$$\text{range} = \{1, 2, 6\}$$

Example 2: Find the domain and range in item (b) of Figure 8-1.

$$\text{domain} = \{1, 2, 3\}$$
$$\text{range} = \{4, 5\}$$

Example 3: Find the domain and range in item (c) of Figure 8-1.

$$\text{Domain } \{-2, -1, 0, 1, 2\}$$
$$\text{Range } \{-2, -1, 1, 3\}$$

Example 4: Find the domain and range in item (d) of Figure 8-1.

The domain and range cannot be listed as in the previous examples. In order to visualize the domain, imagine each point of the graph going vertically to the *x*-axis. The points on the *x*-axis become the domain (see Figure 8-2).

Figure 8-2 The domain is horizontal.

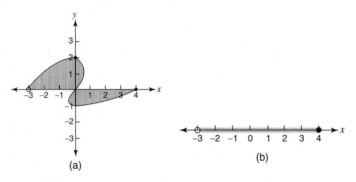

(a)

(b)

The domain can now be expressed as domain = $\{x | -3 < x \le 4\}$, which is read as "the set of *x*'s such that *x* is greater than −3 and *x* is less than or equal to 4."

To visualize the range, have all the points move horizontally to the *y*-axis. The points on the *y*-axis become the range (see Figure 8-3).

Figure 8-3 The range is vertical.

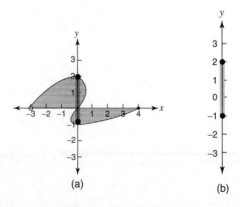

(a)

(b)

The range can be expressed as range = $\{y|-1 \le y \le 2\}$, which is read as "the set of y's such that y is greater than or equal to -1 and y is less than or equal to 2."

Example 5: Find the domain and range in item (e) of Figure 8-1.

Since any value for x produces a y-value and any value for y produces an x-value,

$$\text{domain} = \{\text{all real numbers}\}$$
$$\text{range} = \{\text{all real numbers}\}$$

Function

A relation in which none of the domain values are repeated is called a **function.**

Example 6: Which of the examples given in Figure 8-1 are functions?

(a) $\{(4,1), (3,2), (-1,6)\}$

This is a function, since domain values are not repeated.

(b) The example, shown in Figure 8-4, is not a function.

Figure 8-4 This is not a function.

The diagram represents the set of ordered pairs $\{(1,4), (1,5), (3,4), (2,5)\}$ and $(1,4)$ and $(1,5)$ repeat domain values.

(c) The example, shown in Figure 8-5, is not a function.

Figure 8-5   Like Figure 8-4, this is also not a function.

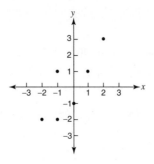

The graph represents the set of ordered pairs {(–2,–2), (–1,–2), (–1,1), (0,–1), (1,1), (2,3)}, and (–1,–2) and (–1,1) repeat domain values.

Notice that the vertical line $x = -1$ would pass through two points of the graph, as shown in Figure 8-6.

Figure 8-6   Functions have no repeated *x* values.

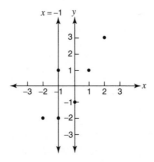

A vertical line that passes through (intersects) a graph in more than one point indicates ordered pairs that have repeated domain values and eliminates the relation from being called a function. This test for functions is called the **vertical line test.**

(d)   The example shown in Figure 8-7 is also not a function. It fails the vertical line test.

Figure 8-7   This fails the vertical line test and is, therefore, not a function.

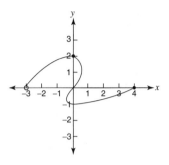

(e) $y = 2x + 3$.

This is a function, since domain values are not repeated. When $y = 2x + 3$ is graphed, it can be seen that the graph passes the vertical line test.

Function Notation

An equation involving x and y and which is also a function can be written in the form $y =$ "some expression involving x"; that is, $y = f(x)$. This last expression is read as "y equals f of x" and means that y is a function of x.

If $y = 4x + 3$, then $f(x) = 4x + 3$. The question "What is y when $x = 2$?" would be expressed as "What is $f(2)$?" if

$$f(x) = 4x + 3$$

then

$$f(2) = 4(2) + 3$$

$$= 11$$

The statement $f(2) = 11$ says, "When x equals 2, y equals 11"; that is, the ordered pair (2, 11) belongs to this function.

Example 7: If $f(x) = x^2 + 2x + 3$, find each of the following.

(a) $f(-3)$ (b) $f(0)$

(a)
$$f(x) = x^2 + 2x + 3$$

$$f(-3) = (-3)^2 + 2(-3) + 3$$

$$= 9 - 6 + 3$$

$$= 6$$

(b)
$$f(x) = x^2 + 2x + 3$$
$$f(0) = (0)^2 + 2(0) + 3$$
$$= 3$$

Compositions of Functions

When the variable name in a function is replaced with another function, the result is called a **composite function.** If

$$f(x) = 3x + 2 \qquad \text{and} \qquad g(x) = 4x - 5$$

then $f[g(x)]$ is a composite function. The statement $f[g(x)]$ is read "f of g of x" or "the composition of f with g." $f[g(x)]$ can also be written as

$$(f \circ g)(x) \qquad \text{or} \qquad f \circ g(x)$$

The symbol between f and g is a small open circle.

Example 8: If $f(x) = 3x + 2$ and $g(x) = 4x - 5$, find each of the following.

(a) $f[g(4)]$ (b) $g \circ f(4)$ (c) $f[g(x)]$ (d) $(g \circ f)(x)$

(a)
$$f(x) = 3x + 2$$
$$f[g(4)] = 3[g(4)] + 2$$
$$= 3(11) + 2$$
$$= 35$$

$$\begin{array}{c} g(x) = 4x - 5 \\ g(4) = 4(4) - 5 \\ = 16 - 5 \\ = 11 \end{array}$$

(b)
$$g(x) = 4x - 5$$
$$g \circ f(4) = 4[f(4)] - 5$$
$$= 4(14) - 5$$
$$= 51$$

$$\begin{array}{c} f(x) = 3x + 2 \\ f(4) = 3(4) + 2 \\ = 12 + 2 \\ = 14 \end{array}$$

(c)
$$f(x) = 3x + 2$$
$$f[g(x)] = 3[g(x)] + 2$$
$$= 3(4x - 5) + 2$$
$$= 12x - 15 + 2$$
$$= 12x - 13$$

$$g(x) = 4x - 5$$

(d) $g(x) = 4x - 5$ $f(x) = 3x + 2$

$(g \circ f)(x) = 4[f(x)] - 5$

$= 4(3x + 2) - 5$

$= 12x + 8 - 5$

$= 12x + 3$

From the previous example, notice that $f[g(x)]$ and $g[f(x)]$ are not equal.

Example 9: If $f(x) = 3x^2 + 2x + 1$ and $g(x) = 4x - 5$, find each of the following.

(a) $f[g(x)]$ (b) $g[f(x)]$

(a) $f(x) = 3x^2 + 2x + 1$ $g(x) = 4x - 5$

$f[g(x)] = 3[g(x)]^2 + 2[g(x)] + 1$

$= 3(4x - 5)^2 + 2(4x - 5) + 1$

$= 3(16x^2 - 40x + 25) + 8x - 10 + 1$

$= 48x^2 - 120x + 75 + 8x - 10 + 1$

$= 48x^2 - 112x + 66$

(b) $g(x) = 4x - 5$ $f(x) = 3x^2 + 2x + 1$

$g[f(x)] = 4[f(x)] - 5$

$= 4(3x^2 + 2x + 1) - 5$

$= 12x^2 + 8x + 4 - 5$

$= 12x^2 + 8x - 1$

Algebra of Functions

If two functions have a common domain, then arithmetic can be performed with them using the following definitions.

$$(f + g)(x) = f(x) + g(x)$$

$$(f - g)(x) = f(x) - g(x)$$

$$(f \cdot g)(x) = f(x)g(x)$$

$$\left(\frac{f}{g}\right)(x) = \frac{f(x)}{g(x)}, \text{ where } g(x) \neq 0$$

Example 10: If $f(x) = x + 4$ and $g(x) = x^2 - 2x - 3$, find each of the following and determine the common domain.

      (a) $(f + g)(x)$ (b) $(f - g)(x)$ (c) $(f \cdot g)(x)$ (d) $(f/g)(x)$

(a)
$$(f + g)(x) = (x + 4) + (x^2 - 2x - 3)$$
$$= x^2 - x + 1$$

The common domain is {all real numbers}.

(b)
$$(f - g)(x) = (x + 4) - (x^2 - 2x - 3)$$
$$= x + 4 - x^2 + 2x + 3$$
$$= -x^2 + 3x + 7$$

The common domain is {all real numbers}.

(c)
$$(f \cdot g)(x) = (x + 4) \cdot (x^2 - 2x - 3)$$

Use the distributive property.

$$= x^3 - 2x^2 - 3x + 4x^2 - 8x - 12$$
$$= x^3 + 2x^2 - 11x - 12$$

The common domain is {all real numbers}.

(d)
$$\left(\frac{f}{g}\right)(x) = \frac{x + 4}{x^2 - 2x - 3}$$
$$= \frac{x + 4}{(x - 3)(x + 1)}$$

This expression is undefined when $x = 3$ or when $x = -1$. So the common domain is {all real numbers except 3 or -1}.

Inverse Functions

If the ordered pairs of a relation R are reversed, then the new set of ordered pairs is called the **inverse relation** of the original relation.

Example 11: If $R = \{(1,2), (3,8), (5,6)\}$, find the inverse relation of R. (The inverse relation of R is written R^{-1}).

$$R^{-1} = \{(2,1), (8,3), (6,5)\}$$

Notice that the domain of R^{-1} is the range of R, and the range of R^{-1} is the domain of R. If a relation and its inverse are graphed, they will be symmetrical about the line $y = x$.

Example 12: Graph R and R^{-1} from Example 11 along with the line $y = x$ on the same set of coordinate axes.

The answer is shown in Figure 8-8.

Figure 8-8   Symmetrical sets of points.

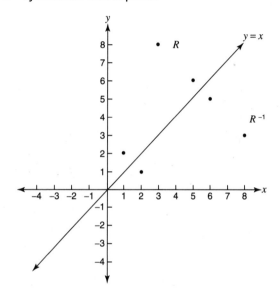

If this graph were "folded over" the line $y = x$, the set of points called R would coincide with the set of points called R^{-1}, making the two sets symmetrical about the line $y = x$.

■ **Identity function.** The function $y = x$, or $f(x) = x$, is called the **identity function**, since for each replacement of x, the result is identical to x.

■ **Inverse function.** Two functions, f and g, are inverses of each other when the composition $f[g(x)]$ and $g[f(x)]$ are both the identity function. That is, $f[g(x)] = g[f(x)] = x$.

Example 13: If $f(x) = 4x - 5$, find $f^{-1}(x)$.

$$f(x) = 4x - 5 \qquad \text{means} \qquad y = 4x - 5$$

To find $f^{-1}(x)$, simply reverse the x and y variables and solve for y.

The original equation is	$y = 4x - 5$
The inverse is	$x = 4y - 5$
Solve for y.	$4y - 5 = x$
	$4y = x + 5$
	$y = \dfrac{x + 5}{4}$
Therefore,	$f^{-1} = \dfrac{x + 5}{4}$

For any ordered pair that makes $f(x) = 4x - 5$ true, the reverse ordered pair will make $f^{-1}(x) = \dfrac{(x+5)}{4}$ true.

To show that $f(x)$ and $f^{-1}(x)$ are truly inverses, show that their compositions both equal the identity function.

$$f\left[f^{-1}(x)\right] = f\left[\frac{x+5}{4}\right] = 4\left(\frac{x+5}{4}\right) - 5 = x + 5 - 5 = x$$

$$f^{-1}\left[f(x)\right] = f^{-1}\left[4x - 5\right] = \frac{(4x-5)+5}{4} = \frac{4x}{4} = x$$

Since $f\left[f^{-1}(x)\right] = f^{-1}\left[f(x)\right] = x$, then $f(x)$ and $f^{-1}(x)$ are inverses of each other.

Example 14: Graph $f(x)$ and $f^{-1}(x)$ from Example 13 together with the identity function on the same set of coordinate axes. The answer is shown in Figure 8-9.

Figure 8-9 Symmetrical graphs.

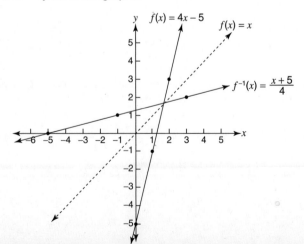

Notice that if the graph were "folded over" the identity function, the graphs of $f(x)$ and $f^{-1}(x)$ would coincide.

Example 15: If $f(x) = x^2$, find $f^{-1}(x)$.

$f(x) = x^2$ means $y = x^2$

The original equation is $\qquad\qquad y = x^2$

The inverse is $\qquad\qquad\qquad x = y^2$

Solve for y. $\qquad\qquad\qquad y^2 = x$

$$y = \pm\sqrt{x}$$

There are two relations for $f^{-1}(x)$,

$$f^{-1}(x) = \sqrt{x}, \qquad\qquad f^{-1}(x) = -\sqrt{x}$$

In order for both $f(x)$ and $f^{-1}(x)$ to be functions, a restriction needs to be made on the domain of $f(x)$ so only one relation appears as $f^{-1}(x)$. If the domain of $f(x)$ is restricted to $\{x | x \geq 0\}$, the $f^{-1}(x) = \sqrt{x}$ is the only answer for $f^{-1}(x)$. If the domain of $f(x)$ is restricted to $\{x | x \leq 0\}$, then $f^{-1}(x) = -\sqrt{x}$ is the only answer for $f^{-1}(x)$.

Example 16: Graph $f(x) - x^2$ together with $f^{-1}(x) = \sqrt{x}, f^{-1}(x) = -\sqrt{x}$, and the identity function $f(x) = x$ all on the same set of coordinate axes.

To graph $f(x) = x^2$, find several ordered pairs that make the sentence $y = x^2$ true. To graph $f^{-1}(x) = \pm\sqrt{x}$, simply take the reverse of the ordered pairs found for $f(x) = x^2$. The graph is as shown in Figure 8-10.

x	$f(x) = x^2$		x	$f^{-1}(x) = \sqrt{x}$		x	$f^{-1}(x) = -\sqrt{x}$
-3	9		9	3		9	-3
-2	4		4	2		4	-2
-1	1		1	1		1	-1
0	0		0	0		0	0
1	1						
2	4						
3	9						

Figure 8-10 $f^{-1}(x)$ is not a function.

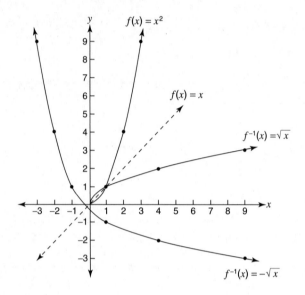

Notice that $f(x) = x^2$ is a function but that $f^{-1}(x) = \pm\sqrt{x}$ is not a function. The reason is that $f^{-1}(x) = \pm\sqrt{x}$ does not pass the vertical line test. Also notice that $f(x)$ and $f^{-1}(x)$ will coincide when the graph is "folded over" the identity function. Thus, the two relations are inverses of each other.

Example 17: Graph $f(x) = x^2$ with the restricted domain $\{x|x \geq 0\}$ together with $f^{-1}(x) = \sqrt{x}$ and the identity function on the same set of coordinate axes. The answer is shown in Figure 8-11.

Figure 8-11   Solution to Example 17.

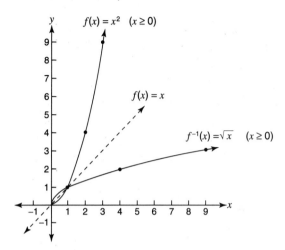

Notice that $f(x)$ and $f^{-1}(x)$ are now both functions, and they are symmetrical with respect to $f(x) = x$. To show that $f(x) = x^2$ and $f^{-1}(x) = \sqrt{x}$ are inverse functions, show that their compositions each produce the identity function.

$$f[f^{-1}(x)] = f[\sqrt{x}] = (\sqrt{x})^2 = x$$
$$f^{-1}[f(x)] = f^{-1}[x^2] = \sqrt{x^2} = x \quad (\text{since } x \geq 0)$$

Chapter Checkout

1. If $f(x) = x^2 + 7x + 5$, find the following

   a. $f(7)$

   b. $f(-3)$

   c. $f(0)$

2. If $f(x) = 3x-7$, find $f^{-1}(x)$

3. If $f(x) = x + 5$ and $g(x) = x^2 - 4x - 5$, find

   a. $(f + g)(x)$

   b. $(f \cdot g)(x)$

Answers: 1. a. 103 b. –7 c. 5   2. $\frac{x+7}{3}$ 3. a. $x^2 - 3x$ b. $x^3 + x^2 - 25x - 25$

Chapter 9

POLYNOMIAL FUNCTIONS

Chapter Check-In

❑ Applying the remainder theorem

❑ Using the factor theorem

❑ Testing for zeros and rational zeros

❑ Graphing polynomial functions

This chapter deals with polynomial functions, showing how to evaluate them and how much can be determined about them by use of the factoring theorem, the remainder theorem, and their zeros. A zero may be any replacement for the variable that will result in the function having a value of zero. Graphing of polynomial functions is also dealt with in detail.

Polynomial Function

A **polynomial function** is any function of the form

$$P(x) = a_0 x^n + a_1 x^{n-1} + a_2 x^{n-2} + \ldots + a_{n-1} x + a_n$$

where the coefficients a_0, a_1, a_2, ..., a_n are real numbers and n is a whole number. Polynomial functions are evaluated by replacing the variable with a value. The instruction "evaluate the polynomial function $P(x)$ when x is replaced with 4" is written as "find $P(4)$."

Example 1: If $P(x) = 3x^3 - 2x^2 + 5x + 3$, find $P(-4)$.

$$P(-4) = 3(-4)^3 - 2(-4)^2 + 5(-4) + 3$$

$$= 3(-64) - 2(16) - 20 + 3$$

$$= -192 - 32 - 20 + 3$$

$$= -241$$

Example 2: If $f(x) = 3x^2 - 4x + 5$, find $f(x + h)$.

$$f(x + h) = 3(x + h)^2 - 4(x + h) + 5$$
$$= 3(x^2 + 2xh + h^2) - 4x - 4h + 5$$
$$= 3x^2 + 6xh + 3h^2 - 4x - 4h + 5$$

Remainder Theorem

If a polynomial $P(x)$ is divided by $(x - r)$, then the remainder of this division is the same as evaluating $P(r)$, and evaluating $P(r)$ for some polynomial $P(x)$ is the same as finding the remainder of $P(x)$ divided by $(x - r)$.

Example 3: Find $P(-3)$ if $P(x) = 7x^5 - 4x^3 + 2x - 11$.

There are two methods of finding $P(-3)$.

■ **Method 1:** Direct replacement of -3 for x

■ **Method 2:** Find the remainder of $P(x)$ divided by $[x - (-3)]$

To use method 1:

$$P(x) = 7x^5 - 4x^3 + 2x - 11$$
$$P(-3) = 7(-3)^5 - 4(-3)^3 + 2(-3) - 11$$
$$= 7(-243) - 4(-27) - 6 - 11$$
$$= -1701 + 108 - 6 - 11$$
$$= -1610$$

By method 2, $P(x)$ divided by $[x - (-3)]$ will be done by synthetic division.

-3	7	0	-4	0	2	-11
		-21	63	-177	531	-1599
	7	-21	59	-177	533	-1610

Therefore, $P(-3) = -1610$.

Example 4: Find the remainder of $P(x)$ divided by $(x - 4)$ if

$$P(x) = x^4 + x^3 - 13x^2 - 25x - 12$$

By method 1,

$$\text{remainder} = P(4)$$
$$= (4)^4 + (4)^3 - 13(4)^2 - 25(4) - 12$$
$$= 256 + 64 - 208 - 100 - 12$$
$$= 0$$

By method 2, synthetic division,

$$
\underline{4|} \quad
\begin{array}{ccccc}
1 & 1 & -13 & -25 & -12 \\
  & 4 & 20 & 28 & 12 \\
\hline
1 & 5 & 7 & 3 & 0
\end{array}
$$

Therefore, the remainder = 0.

In Example 4, since the division has a remainder of zero, both the **divisor** (the number doing the dividing) and the **quotient** (the answer) are **factors** of the **dividend** (the number being divided).

Factor Theorem

If $P(x)$ is a polynomial, then $P(r) = 0$ if and only if $x - r$ is a factor of $P(x)$.

Example 5: Is $(x + 2)$ a factor of $x^3 - x^2 - 10x - 8$?

Check to see if $(x^3 - x^2 - 10x - 8) \div (x + 2)$ has a remainder of zero. Using synthetic division, you get

$$
\underline{-2|} \quad
\begin{array}{cccc}
1 & -1 & -10 & -8 \\
  & -2 & 6 & 8 \\
\hline
1 & -3 & -4 & \boxed{0}
\end{array}
$$
⟵ remainder

Because the remainder of the division is zero, $(x + 2)$ is a factor of $x^3 - x^2 - 10x - 8$. The expression $x^3 - x^2 - 10x - 8$ can now be expressed in factored form.

$$x^3 - x^2 - 10x - 8 = (x + 2)(x^2 - 3x - 4)$$

But $(x^2 - 3x - 4)$ can be factored further into $(x - 4)(x + 1)$. So

$$x^3 - x^2 - 10x - 8 = (x + 2)(x - 4)(x + 1)$$

The expression $x^3 - x^2 - 10x - 8$ is now **completely factored.** From this form, it is also seen that $(x - 4)$ and $(x + 1)$ are also factors of $x^3 - x^2 - 10x - 8$.

Zeros of a Function

The **zero of a function** is any replacement for the variable that will produce an answer of zero. Graphically, the real zero of a function is where the graph of the function crosses the x-axis; that is, the real zero of a function is the x-intercept(s) of the graph of the function.

Example 6: Find the zeros of the function $f(x) = x^2 - 8x - 9$.

Find x so that $f(x) = x^2 - 8x - 9 = 0$. $f(x)$ can be factored, so begin there.

$$f(x) = x^2 - 8x - 9 = 0$$
$$(x + 1)(x - 9) = 0$$

$$x + 1 = 0 \quad \text{or} \quad x - 9 = 0 \qquad \text{(zero product rule)}$$
$$x = -1 \qquad\qquad x = 9$$

Therefore, the zeros of the function $f(x) = x^2 - 8x - 9$ are -1 and 9. This means

$$f(-1) = 0 \quad \text{and} \quad f(9) = 0$$

If a polynomial function with integer coefficients has real zeros, then they are either rational or irrational values. Rational zeros can be found by using the rational zero theorem.

Rational Zero Theorem

If a polynomial function, written in descending order, has integer coefficients, then any rational zero must be of the form $\pm\, p/q$, where p is a factor of the constant term and q is a factor of the leading coefficient.

Example 7: Find all the rational zeros of

$$f(x) = 2x^3 + 3x^2 - 8x + 3$$

According to the rational zero theorem, any rational zero must have a factor of 3 in the numerator and a factor of 2 in the denominator.

$$p: \quad \text{factors of } 3 = \pm 1, \pm 3$$
$$q: \quad \text{factors of } 2 = \pm 1, \pm 2$$

The possibilities of p/q, in simplest form, are

$$\pm 1, \quad \pm \tfrac{1}{2}, \quad \pm 3, \quad \pm \tfrac{3}{2}$$

These values can be tested by using direct substitution or by using synthetic division and finding the remainder. Synthetic division is the better method because if a zero is found, the polynomial can be written in factored form and, if possible, can be factored further, using more traditional methods.

Example 8: Find rational zeros of $2x^3 + 3x^2 - 8x + 3$ by using synthetic division.

p/q	2	3	-8	3	
1	2	5	-3	0	1 is a zero
-1	2	1	-9	12	
$\frac{1}{2}$	2	4	-6	0	$\frac{1}{2}$ is a zero
$-\frac{1}{2}$	2	2	-9	$\frac{15}{2}$	
3	2	9	19	60	
-3	2	-3	1	0	-3 is a zero
$\frac{3}{2}$	2	6	1	$\frac{9}{2}$	
$-\frac{3}{2}$	2	0	-8	15	

The zeros of $f(x) = 2x^3 + 3x^2 - 8x + 3$ are 1, $\frac{1}{2}$, and -3. This means

$$f(1) = 0, \qquad f(\tfrac{1}{2}) = 0, \quad \text{and} \qquad f(-3) = 0$$

The zeros could have been found without doing so much synthetic division. From the first line of the chart, 1 is seen to be a zero. This allows $f(x)$ to be written in factored form using the synthetic division result.

$$f(x) = 2x^3 + 3x^2 - 8x + 3 = (x - 1)(2x^2 + 5x - 3)$$

But $2x^2 + 5x - 3$ can be further factored into $(2x - 1)(x + 3)$ using the more traditional methods of factoring.

$$2x^2 + 5x - 3 = (x - 1)(2x - 1)(x + 3)$$

From this completely factored form, the zeros are quickly recognized. Zeros will occur when

$x - 1 = 0$	$2x - 1 = 0$	$x + 3 = 0$
$x = 1$	$x = \frac{1}{2}$	$x = -3$

Graphing Polynomial Functions

Polynomial functions of the form $f(x) = x^n$ (where n is a positive integer) form one of two basic graphs, shown in Figure 9-1.

Figure 9-1   Graphs of polynomials.

$f(x) = x^n$, n an even integer $f(x) = x^n$, n an odd integer

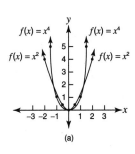

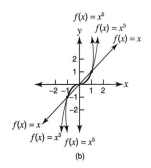

Each graph has the origin as its only x-intercept and y-intercept. Each graph contains the ordered pair $(1, 1)$. If a polynomial function can be factored, its x-intercepts can be immediately found. Then a study is made as to what happens between these intercepts, to the left of the far left intercept and to the right of the far right intercept.

Example 9: Graph $f(x) = x^4 - 10x^2 + 9$.

$f(x) = x^4 - 10x^2 + 9$ can be factored

$$f(x) = x^4 - 10x^2 + 9$$
$$= (x^2 - 1)(x^2 - 9)$$
$$= (x + 1)(x - 1)(x + 3)(x - 3)$$

The zeros of this function are -1, 1, -3, and 3. That is, -1, 1, -3, and 3 are the x-intercepts of this function.

When $x < -3$, say, $x = -4$, then

$$f(x) = (x + 1)(x - 1)(x + 3)(x - 3)$$
$$f(-4) = (-4 + 1)(-4 - 1)(-4 + 3)(-4 - 3)$$
$$= (-3)(-5)(-1)(-7)$$
$$= 105$$

So for $x < -3$, $f(x) > 0$.

When $-1 < x < 1$, say, $x = 0$, then

$$f(x) = (x + 1)(x - 1)(x + 3)(x - 3)$$
$$f(0) = (0 + 1)(0 - 1)(0 + 3)(0 - 3)$$
$$= (1)(-1)(3)(-3)$$
$$= 9$$

So for $-1 < x < 1$, $f(x) > 0$.

In a similar way, it can be seen that

when $x > 3$, $f(x) > 0$
when $-3 < x < -1$, $f(x) < 0$
when $1 < x < 3$, $f(x) < 0$

The graph then has points in the shaded regions as shown in Figure 9-2.

Figure 9-2   Zeros of a function.

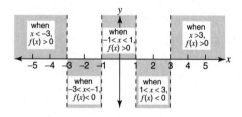

The *y*-intercept of this function is found by finding $f(0)$.

$$f(0) = 9$$

so $(0, 9)$ is a point on the graph. To complete the graph, find and plot several points. Evaluate $f(x)$ for several integer replacements; then connect these points to form a smooth curve (see Figure 9-3).

Figure 9-3 Graph of $f(x)$.

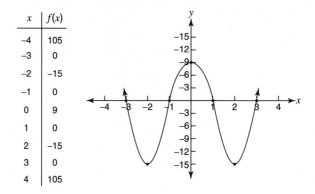

x	f(x)
-4	105
-3	0
-2	-15
-1	0
0	9
1	0
2	-15
3	0
4	105

Notice that $f(x) = x^4 - 10x^2 + 9$ has a leading term with an even exponent. The far right and far left sides of the graph will go in the same direction. Because the leading coefficient is positive, the two sides will go up. If the leading coefficient were negative, the two sides would go down.

Example 10: Graph $f(x) = x^3 - 19x + 30$.

$f(x) = x^3 - 19x + 30$ can be factored using the rational zero theorem:

p/q	1	0	-19	30	
1	1	1	-18	12	
-1	1	-1	-18	48	
2	1	2	-15	0	2 is a zero of the function

$f(x)$ can now be written in factored form and further factored.

$$f(x) = x^3 - 19x + 30$$
$$= (x - 2)(x^2 + 2x - 15)$$
$$= (x - 2)(x - 3)(x + 5)$$

The zeros of this function are 2, 3, and –5 (see Figure 9-4).

Figure 9-4 A cubic equation.

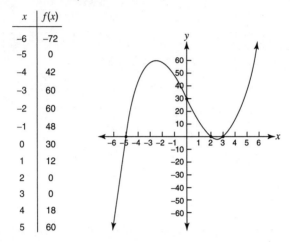

x	$f(x)$
−6	−72
−5	0
−4	42
−3	60
−2	60
−1	48
0	30
1	12
2	0
3	0
4	18
5	60

Notice that $f(x) = x^3 - 19x + 30$ has a leading term that has a positive coefficient and an odd exponent. This function will always go up toward the far right and down toward the far left. If the leading coefficient were negative with an odd exponent, the graph would go up toward the far left and down toward the far right.

Chapter Checkout

1. Find $P(-2)$ if $P(x) = 6x^4 - 5x^3 + 2x - 12$.

2. Find the zeros of the following function:
$$f(x) = x^2 - x - 12$$

3. Find all rational zeros of the following function:
$$f(x) = 2x^3 + 7x^2 + 2x - 3$$

Answers: 1. 120   2. −3, 4   3. −3, −1, $\frac{1}{2}$

Chapter 10

RADICALS AND COMPLEX NUMBERS

Radicals are the undoing of exponents. In other words, since 2 squared is 4, radical 4 is 2. The radical sign, $\sqrt{}$, is used to indicate "the root" of the number beneath it. If the radical sign is unmodified by a number, then it is the square root of the number under the sign that is being sought. This chapter is concerned with the terminology and operations involving radical expressions. When operating arithmetically on radicals, the radical behaves as if it were a variable.

Complex numbers are natural partners of radicals. They result from finding the radical of a negative expression. The imaginary number, i, is used to represent the square root of 1. So, $3i$ expresses the simplification of $\sqrt{-9}$.

Radicals

The expression $\sqrt[n]{a}$ is called a **radical expression.** The symbol $\sqrt{}$ is called the **radical sign.** The expression under the radical sign is called the **radicand,** and n, an integer greater than 1, is called the **index.** If the radical expression appears without an index, the index is assumed to be 2. The expression $\sqrt[n]{a}$ is read as "the nth root of a." Remember:

$$\text{If } x = \sqrt[n]{a}, \text{ then } x^n = a$$

Example 1: Simplify each of the following.

(a) $\sqrt{25}$ (d) $\sqrt[3]{-\frac{1}{8}}$

(b) $\sqrt[3]{-\frac{1}{8}}$ (e) $\sqrt{-4}$

(c) $\sqrt[5]{\frac{1}{32}}$

(a) If $x = \sqrt{25}$, then $x^2 = 25$.

$$x = 5 \quad \text{or} \quad x = -5$$

Because x could be either value, a rule is established. If a radical expression could have either a positive or a negative answer, then you always take the positive. This is called the "principal root." Thus,

$$\sqrt{25} = 5 \quad \text{and} \quad -\sqrt{25} = -5$$

(b) If $x = \sqrt[3]{64}$, then

$$x^3 = 64 \quad \text{and} \quad x = 4$$

so $\sqrt[3]{64} = 4$

(c) If $x = \sqrt[5]{\frac{1}{32}}$, then

$$x^5 = \frac{1}{32} \quad \text{and} \quad x = \frac{1}{2}$$

so $\sqrt[5]{\frac{1}{32}} = \frac{1}{2}$

(d) If $x = \sqrt[3]{-\frac{1}{8}}$, then

$$x^3 = -\frac{1}{8} \quad \text{and} \quad x = -\frac{1}{2}$$

so $\sqrt[3]{-\frac{1}{8}} = -\frac{1}{2}$

(e) If $x = \sqrt{-4}$, then $x^2 = -4$. There is no real value for x, so $\sqrt{-4}$ is not a real number.

Following are true statements regarding radical expressions.

n	a	$\sqrt[n]{a}$	Example
even	positive	positive	$\sqrt[4]{16} = 2$
	negative	not real	$\sqrt[4]{-16}$ not real
	zero	zero	$\sqrt[4]{0} = 0$
odd	positive	positive	$\sqrt[3]{8} = 2$
	negative	negative	$\sqrt[3]{-8} = -2$
	zero	zero	$\sqrt[3]{0} = 0$

When variables are involved, absolute value signs are sometimes needed.

Example 2: Simplify $\sqrt{x^2}$.

It would seem that $\sqrt{x^2} = x$. But there is no guarantee that x is nonnegative. Because of this, $\sqrt{x^2}$ is expressed as $|x|$, which guarantees that the result is nonnegative.

Absolute value signs are *never* used when the index is odd. Absolute value signs are sometimes used when the index is even, at those times when the result could possibly be negative.

Example 3: Simplify the following, using absolute value signs when needed.

$$\text{(a) } \sqrt[4]{16x^8 y^{24}} \quad \text{(b) } \sqrt[4]{16x^8 y^{20}} \quad \text{(c) } \sqrt[5]{32x^{10} y^{15}}$$

(a) $\sqrt[4]{16x^8 y^{24}} = \sqrt[4]{2^4 \left(x^2\right)^4 \left(y^6\right)^4} = \left|2x^2 y^6\right|$

Since $2x^2y^6$ could not be negative even if x or y were negative, absolute value signs are not needed.

(b) $\sqrt[4]{16x^8 y^{20}} = \sqrt[4]{2^4 \left(x^2\right)^4 \left(y^5\right)^4} = \left|2x^2 y^5\right|$

Since the expression could be negative if y were negative, a correct way to represent the answer is $|2x^2y^5|$. Because only y could have caused the answer to be negative, another way to represent the answer is $2x^2|y|^5$.

(c) $\sqrt[5]{32x^{10} y^{15}} = \sqrt[5]{2^5 \left(x^2\right)^5 \left(y^3\right)^5} = 2x^2 y^3$

Absolute value signs are never used when the index is odd.

Simplifying Radicals

Simplifying radicals involves the product rule.

Product Rule for Radicals

If n is even, and $a \geq 0$, $b \geq 0$, then

$$\sqrt[n]{ab} = \sqrt[n]{a}\,\sqrt[n]{b}$$

If n is odd, then for all real numbers a and b,

$$\sqrt[n]{ab} = \sqrt[n]{a}\,\sqrt[n]{b}$$

Example 4: Simplify each of the following.

$$\text{(a) } \sqrt{20} \qquad\qquad \text{(c) } \sqrt{(x-3)^6}$$
$$\text{(b) } \sqrt[3]{64x^9\,y^{10}\,z^4} \qquad \text{(d) } \sqrt[3]{(y-5)^5}$$

(a) $\sqrt{20} = \sqrt{2^2 \cdot 5} = 2\sqrt{5}$

(b) $\sqrt[3]{64x^9\,y^{10}\,z^4} = \sqrt[3]{(4)^3 \left(x^3\right)^3 \left(y^3\right)^3 z^3\,yz} = 4x^3\,y^3\,z\,\sqrt[3]{yz}$

(c) $\sqrt{(x-3)^6} = \sqrt{\left[(x-3)^3\right]^2} = \left|(x-3)^3\right|$ or $|x-3|^3$

(d) $\sqrt[3]{(y-5)^5} = \sqrt[3]{(y-5)^3 (y-5)^2} = (y-5)\sqrt[3]{(y-5)^2}$

Example 5: Simplify $\sqrt[3]{12x^2} \cdot \sqrt[3]{4x^4}$

$$\sqrt[3]{12x^2} \cdot \sqrt[3]{4x^4} = \sqrt[3]{\left(12x^2\right)\left(4x^4\right)}$$
$$= \sqrt[3]{48x^6}$$
$$= \sqrt[3]{2^3 \cdot 2 \cdot 3 \left(x^2\right)^3}$$
$$= 2x^2 \sqrt[3]{6}$$

Adding and Subtracting Radical Expressions

Radical expressions are called **like radical expressions** if the indexes are the same and the radicands are identical. $\sqrt[3]{5x}$ and $4\sqrt[3]{5x}$ are like radical expressions, since the indexes are the same and the radicands are identical, but $\sqrt[3]{5x}$ and $\sqrt[3]{5y}$ are not like radical expressions, since their radicands are not identical. Radical expressions can be added or subtracted only if they are *like radical expressions.*

Example 6: Simplify each of the following.

$$\text{(a) } 4 + 3\sqrt{7} + 6 + 5\sqrt{7}$$
$$\text{(b) } 5\sqrt{27} + 6\sqrt{3} - 4\sqrt{48}$$
$$\text{(c) } \sqrt[3]{24a} + \sqrt[3]{81a}$$

(a) $\quad 4 + 3\sqrt{7} + 6 + 5\sqrt{7} = (4 + 6) + \left(3\sqrt{7} + 5\sqrt{7}\right)$

$$= 10 + 8\sqrt{7}$$

(b) $\quad 5\sqrt{27} + 6\sqrt{3} - 4\sqrt{48} = 5\sqrt{3^2 \cdot 3} + 6\sqrt{3} - 4\sqrt{(4)^2 \cdot 3}$

$$= 5 \cdot 3\sqrt{3} + 6\sqrt{3} - 4 \cdot 2^2\sqrt{3}$$
$$= 15\sqrt{3} + 6\sqrt{3} - 16\sqrt{3}$$
$$= 5\sqrt{3}$$

(c) $\quad \sqrt[3]{24a} + \sqrt[3]{81a} = \sqrt[3]{2^3 \cdot 3a} + \sqrt[3]{3^3 \cdot 3a}$

$$= 2\sqrt[3]{3a} + 3\sqrt[3]{3a}$$
$$= 5\sqrt[3]{3a}$$

Multiplying Radical Expressions

To multiply radical expressions, use the distributive property and the product rule for radicals.

Example 7: Simplify each of the following.

$$\text{(a) } \left(4 + 3\sqrt{2}\right)\left(\sqrt{10} + \sqrt{5}\right) \quad \text{(b) } \sqrt{6}\left(\sqrt{3} + \sqrt{12}\right)$$

(a) $\quad \left(4 + 3\sqrt{2}\right)\left(\sqrt{10} + \sqrt{5}\right) = 4\sqrt{10} + 4\sqrt{5} + 3\sqrt{2}\sqrt{10} + 3\sqrt{2}\sqrt{5}$

$$= 4\sqrt{10} + 4\sqrt{5} + 3\sqrt{20} + 3\sqrt{10}$$
$$= 4\sqrt{10} + 4\sqrt{5} + 6\sqrt{5} + 3\sqrt{10}$$
$$= 7\sqrt{10} + 10\sqrt{5}$$

(b) $\sqrt{6}\left(\sqrt{3}+\sqrt{12}\right)=\sqrt{6}\sqrt{3}+\sqrt{6}\sqrt{12}$

$$=\sqrt{18}+\sqrt{72}$$
$$=3\sqrt{2}+6\sqrt{2}$$
$$=9\sqrt{2}$$

Dividing Radical Expressions

When dividing radical expressions, use the quotient rule.

> *Quotient Rule for Radicals*
>
> For all real values, a and b, $b \neq 0$
>
> 1. If n is even, and $a \geq 0$, $b > 0$, then
>
> $$\sqrt[n]{\frac{a}{b}} = \frac{\sqrt[n]{a}}{\sqrt[n]{b}}$$
>
> 2. If n is odd, and $b \neq 0$, then
>
> $$\sqrt[n]{\frac{a}{b}} = \frac{\sqrt[n]{a}}{\sqrt[n]{b}}$$

Radical expressions are written in simplest terms when

- The index is as small as possible.

- The radicand contains no factor (other than 1) which is the nth power of an integer or polynomial.

- The radicand contains no fractions.

- No radicals appear in the denominator.

Example 8: Simplify each of the following.

(a) $\sqrt[3]{\frac{3}{8}}$ (b) $\frac{4\sqrt{14}}{8\sqrt{2}}$

(a) Using the quotient rule for radicals, $\sqrt[3]{\dfrac{3}{8}} = \dfrac{\sqrt[3]{3}}{\sqrt[3]{8}}$

$$= \dfrac{\sqrt[3]{3}}{2}$$

(b) Using the quotient rule for radicals, $\dfrac{4\sqrt{14}}{8\sqrt{2}}$

$$= \dfrac{4}{8} \cdot \sqrt{\dfrac{14}{2}}$$

$$= \dfrac{1}{2}\sqrt{7}$$

$$= \dfrac{\sqrt{7}}{2}$$

Rationalizing the denominator

An expression with a radical in its denominator should be simplified into one without a radical in its denominator. This process is called **rationalizing the denominator.** This is accomplished by multiplying the expression by the value 1 in an appropriate form.

Example 9: Simplify each of the following.

$$\text{(a) } \sqrt{\dfrac{7}{x}} \qquad\qquad \text{(b) } \sqrt[3]{\dfrac{12x^2}{5y^2}}$$

(a) Rationalizing the denominator, $\sqrt{\dfrac{7}{x}} = \dfrac{\sqrt{7}}{\sqrt{x}} \cdot \dfrac{\sqrt{x}}{\sqrt{x}}$

$$= \dfrac{\sqrt{7x}}{x}$$

(b) $\sqrt[3]{\dfrac{12x^2}{5y^2}} = \dfrac{\sqrt[3]{12x^2}}{\sqrt[3]{5y^2}}$

What can be multiplied with $\sqrt[3]{5y^2}$ so the result will not involve a radical? The answer is $\sqrt[3]{5^2\,y}$ or $\sqrt[3]{25y}$. Therefore,

$$\sqrt[3]{\dfrac{12x^2}{5y^2}} = \dfrac{\sqrt[3]{12x^2}}{\sqrt[3]{5y^2}} \cdot \dfrac{\sqrt[3]{25y}}{\sqrt[3]{25y}}$$

$$= \dfrac{\sqrt[3]{300x^2\,y}}{5y}$$

Conjugates

If a and b are unlike terms, then the **conjugate** of $a + b$ is $a - b$, and the conjugate of $a - b$ is $a + b$. The conjugate of $(5 + \sqrt{3})$ is $(5 - \sqrt{3})$. Conjugates are used for rationalizing the denominator when the denominator is a two-termed expression involving a square root.

Example 10: Simplify $\dfrac{6 + \sqrt{3}}{4 - \sqrt{3}}$.

To rationalize the denominator of this expression, multiply by one in the form of the conjugate over itself.

$$\frac{6 + \sqrt{3}}{4 - \sqrt{3}} = \frac{6 + \sqrt{3}}{4 - \sqrt{3}} \cdot \frac{\left(4 + \sqrt{3}\right)}{\left(4 + \sqrt{3}\right)}$$

$$= \frac{24 + 6\sqrt{3} + 4\sqrt{3} + 3}{16 + 4\sqrt{3} - 4\sqrt{3} - 3}$$

$$= \frac{27 + 10\sqrt{3}}{13}$$

Rational Exponents

If n is a natural number greater than 1, and b is any nonnegative real number, then

$$b^{1/n} = \sqrt[n]{b}$$

and since a negative exponent indicates a reciprocal, then

$$b^{-(1/n)} = \frac{1}{\sqrt[n]{b}} \qquad \text{when } b \neq 0$$

Example 11: Simplify each of the following.

(a) $36^{1/2}$ (b) $(72x^4y)^{1/3}$ (c) $16^{-(1/4)}$

(a) $36^{1/2} = \sqrt{36} = 6$

(b) $(72x^4y)^{1/3} = \sqrt[3]{72x^4\, y} = 2x\sqrt[3]{9xy}$

(c) $16^{-(1/4)} = \dfrac{1}{\sqrt[4]{16}} = \dfrac{1}{2}$

If n is a natural number greater than 1, m is an integer, and b is a non-negative real number, then

$$b^{m/n} = \left(\sqrt[n]{b}\right)^m$$

and

$$\left(\sqrt[n]{b}\right)^m = \sqrt[n]{b^m}$$

Example 12: Simplify each of the following.

(a) $27^{2/3}$ (b) $25^{-(3/2)}$

(a) $27^{2/3} = \sqrt[3]{27^2} = \left(\sqrt[3]{27}\right)^2 = 3^2 = 9$

(b) $25^{-(3/2)} = \sqrt{25^{-3}} = \left(\sqrt{25}\right)^{-3} = 5^{-3} = \frac{1}{125}$

Complex Numbers

The expression $\sqrt{-1}$ has no real answer. The symbol i is created to represent $\sqrt{-1}$ and is called an **imaginary value.** Since $i = \sqrt{-1}$, $i^2 = -1$. Any expression that is a product of a real number with i is called a **pure imaginary number.**

Example 13: Simplify each of the following.

(a) $\sqrt{-25}$ (b) $\sqrt{-40}$ (c) $(6i)(4i)$ (d) $\sqrt{-6}\sqrt{-8}$

(a) $\sqrt{-25} = \sqrt{25}\sqrt{-1} = 5i$

(b) $\sqrt{-40} = \sqrt{40}\sqrt{-1} = 2\sqrt{10}\cdot i$

This last expression is commonly written as $2i\sqrt{10}$ so that the i is not mistakenly written under the radical.

(c) $(6i)(4i) = 24i^2 = 24(-1) = -24$

(d) $\sqrt{-6}\sqrt{-8} = \left(i\sqrt{6}\right)\left(i\sqrt{8}\right) = i^2\sqrt{48} = -1\left(4\sqrt{3}\right) = -4\sqrt{3}$

For this last example, all imaginary values had to be put into their "i-form" before any simplifying could be done. Note that

$$\sqrt{-6}\sqrt{-8} \neq \sqrt{(-6)(-8)}$$
$$\left(i\sqrt{6}\right)\left(i\sqrt{8}\right) \neq \sqrt{48}$$
$$i^2\sqrt{48} \neq \sqrt{48}$$
$$-1\sqrt{48} \neq \sqrt{48}$$

That is, the product rule for radicals does not hold (in general) with imaginary numbers.

When i is raised to powers, it has a repeating pattern.

$i^0 = 1$	$i^4 = 1$	$i^8 = 1$	$i^{12} = 1$
$i^1 = i$	$i^5 = i$	$i^9 = i$	$i^{13} = i$
$i^2 = -1$	$i^6 = -1$	$i^{10} = -1$	$i^{14} = -1$
$i^3 = -i$	$i^7 = -i$	$i^{11} = -i$	$i^{15} = -i$

When i is raised to any whole number power, the result is always 1, i, -1 or $-i$. If the exponent on i is divided by 4, the remainder will indicate which of the four values is the result.

Example 14: Simplify each of the following.

(a) i^{34} (b) i^{95} (c) i^{108} (d) i^{53}

(a) i^{34}

Since 34 divided by 4 has a remainder of 2, $i^{34} = i^2 = -1$.

(b) i^{95}

Since 95 divided by 4 has a remainder of 3, $i^{95} = i^3 = -i$.

(c) i^{108}

Since 108 divided by 4 has a zero remainder, $i^{108} = i^0 = 1$.

(d) i^{53}

Since 53 divided by 4 has a remainder of 1, $i^{53} = i^1 = i$.

Complex numbers and complex conjugates. A **complex number** is any expression that is a sum of a pure imaginary number and a real number. A complex number is usually expressed in a form called the $a + bi$ form, or standard form, where a and b are real numbers. The expressions $a + bi$ and $a - bi$ are called **complex conjugates.** Complex conjugates are used to rationalize the denominator when dividing with complex numbers.

Arithmetic with complex numbers is done in a similar manner as arithmetic with polynomials. The following are definitions for arithmetic with two complex numbers call $(a + bi)$ and $(c + di)$.

■ Combining like terms and factoring out the i,

$$(a + bi) + (c + di) = (a + c) + (b + d)i$$
$$(a + bi) - (c + di) = (a - c) + (b - d)i$$

■ Using the distributive property,

$$(a + bi)(c + di) = ac + adi + bci + bdi^2$$
$$= ac + adi + bci - bd$$
$$= (ac - bd) + (ad + bc)i$$

■ Rationalizing the denominator,

$$\frac{a + bi}{c + di} = \frac{a + bi}{c + di} \cdot \frac{c - di}{c - di} = \frac{(ac + bd) + (bc - ad)i}{c^2 + d^2}$$

$$\underbrace{\qquad\qquad\qquad}_{\substack{\text{rationalizing} \\ \text{the denominator}}}$$

Example 15: Find the sum, difference, product, and quotient of $(4 + 3i)$ and $(5 - 4i)$.

Sum:
$$(4 + 3i) + (5 - 4i) = (4 + 5) + (3 - 4)i$$
$$= 9 - i$$

Difference:
$$(4 + 3i) - (5 - 4i) = (4 - 5) + [3 - (-4)]i$$
$$= -1 + 7i$$

Product:
$$(4 + 3i)(5 - 4i) = 20 - 16i + 15i - 12i^2$$
$$= 20 - i + 12$$
$$= 32 - i$$

Quotient: Rationalize the denominator.

$$\frac{4+3i}{5-4i} = \frac{4+3i}{5-4i} \cdot \frac{5+4i}{5+4i}$$

$$= \frac{(4 \cdot 5 - 3 \cdot 4) + (3 \cdot 5 + 4 \cdot 4)i}{5^2 + 4^2}$$

$$= \frac{8+31i}{41}$$

$$= \tfrac{8}{41} + \tfrac{31}{41}i$$

Example 16: Simplify $\frac{5}{6i}$.

Since $6i$ is $0 + 6i$, its complex conjugate is $0 - 6i$. Therefore,

$$\frac{5}{6i} = \frac{5}{6i} \cdot \frac{-6i}{-6i}$$

$$= \frac{-30i}{-36i^2}$$

$$= \frac{-30i}{36}$$

$$= -\frac{5}{6}i$$

Chapter Checkout

1. Simplify each of the following:

a) $\sqrt{81}$ b) $\sqrt[4]{256}$ c) $\sqrt[3]{216}$ d) $\sqrt{-16}$

2. Simplify each of the following:

a) $\sqrt{24}$ b) $\sqrt{162}$ c) $\sqrt[3]{8c^6 d^8 f^9}$ d) $27^{-\frac{1}{3}}$

3. Find the sum, difference, product, and quotient of $(6 + 3i)$ and $(4 - 2i)$ in that order. Rationalize any denominators if appropriate.

Answers: 1. a. 9 b. 4 c. 6 d. $4i$ 2. a. $9\sqrt{2}$ c. $2c^2 d^2 f^3 \sqrt[3]{d^2}$ d. $\frac{1}{3}$
3. Sum: $10 + 1$, Difference: $2 + 5i$, Product: $30 - 40i$, Quotient: $\frac{9}{10} + \frac{6}{5}i$

Chapter 11

QUADRATICS IN ONE VARIABLE

Chapter Check-In

- ❑ Solving quadratics by factoring
- ❑ Solving quadratics by completing the square
- ❑ Solving quadratics by formula
- ❑ Solving quadratic equations and inequalities
- ❑ Solving radical equations

The quadratic equation is arguably the signature entity of Algebra II. Everything you have learned up to this point has been preparing you to solve quadratic equations and inequalities. This chapter takes you through the logical applications of testing quadratic expressions for factorability and then, if the expression passes the test, solving by factoring.

Should a quadratic not meet the test of factorability, alternative solutions are developed and applied. When all else has failed, there is the quadratic formula, which unlocks the roots of even the most stubborn of quadratics.

Quadratic Equations

A **quadratic equation** is any equation of the form

$$ax + bx + c = 0 \qquad (a \neq 0)$$

A quadratic equation is usually solved in one of four algebraic ways:

- ■ Factoring
- ■ Applying the square root property
- ■ Completing the square
- ■ Applying the quadratic formula

Solving Quadratics by Factoring

Solving a quadratic equation by factoring depends on the zero product property. The zero product property states that if $ab = 0$, then either $a = 0$ or $b = 0$.

Example 1: Solve $2x^2 = -9x - 4$ by using factoring.

1. **First, get all terms on one side of the equation.**
$$2x^2 = -9x - 4$$
$$2x^2 + 9x + 4 = 0$$

2. **Factor the quadratic.**
$$(2x + 1)(x + 4) = 0$$

3. **Apply the zero product property.**

$$2x + 1 = 0 \qquad \text{or} \qquad x + 4 = 0$$
$$x = -\frac{1}{2} \qquad\qquad\qquad x = -4$$

Solving Quadratics by the Square Root Property

The square root property says that if $x^2 = c$, then $x = \sqrt{c}$ or $x = -\sqrt{c}$. This can be written as "if $x^2 = c$, then $x = \pm\sqrt{c}$." If c is positive, then x has two real answers. If c is negative, then x has two imaginary answers.

Example 2: Solve each of the following equations.

(a) $x^2 = 48$ (d) $(x - 7)^2 = 81$
(b) $x^2 = -16$ (e) $(x + 3)^2 = 24$
(c) $5x^2 - 45 = 0$

$$
\begin{array}{lll}
\text{(a)} \ x^2 = 48 & \text{(b)} \ x^2 = -16 & \text{(c)} \ 5x^2 - 45 = 0 \\
x = \pm\sqrt{48} & x = \pm\sqrt{-16} & 5x^2 = 45 \\
= \pm 4\sqrt{3} & = \pm 4i & x^2 = 9 \\
& & x = \pm\sqrt{9} \\
& & = \pm 3
\end{array}
$$

$$(d)(x-7)^2 = 81 \qquad (e)(x+3)^2 = 24$$

$$(x-7) = \pm\sqrt{81} \qquad (x+3) = \pm\sqrt{24}$$

$$x - 7 = \pm 9 \qquad x + 3 = \pm 2\sqrt{6}$$

$$x = 7 \pm 9 \qquad x = -3 \pm 2\sqrt{6}$$

$$x = 16 \ \text{ or } \ x = -2 \qquad x = -3 + 2\sqrt{6} \ \text{ or } \ x = -3 - 2\sqrt{6}$$

Solving Quadratics by Completing the Square

The expression $x^2 + bx$ can be made into a square trinomial by adding to it a certain value. This value is found by performing two steps:

1. **Multiply b (the coefficient of the "x-term") by $\frac{1}{2}$.**
2. **Square the result.**

Example 3: Find the value to add to $x^2 + 8x$ to make it become a square trinomial.

$$x^2 + 8x$$

Multiply the coefficient of the "x-term" by $\frac{1}{2}$.

$$8(\tfrac{1}{2}) = 4$$

Square that result.

$$(4)^2 = 16$$

So 16 must be added to $x^2 + 8x$ to make it a square trinomial.

$$x^2 + 8x + 16 = (x + 4)^2$$

Finding the value that makes a quadratic become a square trinomial is called **completing the square.**

Example 4: Solve the equation $x^2 - 10x = -16$ by using the completing the square method.

$$x^2 - 10x = -16$$

Multiply coefficient of "x-term" by $\frac{1}{2}$.

$$(-10)(\tfrac{1}{2}) = -5$$

Square the result.

$$(-5)^2 = 25$$

Add 25 to both sides of the equation.

$$x^2 - 10x + 25 = -16 + 25$$
$$(x - 5)^2 = 9$$
$$x - 5 = \pm\sqrt{9}$$
$$x - 5 = \pm 3$$
$$x = 5 \pm 3$$
$$x = 8 \qquad \text{or} \qquad x = 2$$

To solve quadratic equations by using the completing the square method, the coefficient of the squared term must be 1. If it isn't, then first divide both sides of the equation by that coefficient and then proceed as before.

Example 5: Solve $2x^2 - 3x + 4 = 0$ by using the completing the square method.

$$2x^2 - 3x + 4 = 0$$

Get the coefficient of the squared term to be 1.

$$\frac{2x^2}{2} - \frac{3x}{2} + \frac{4}{2} = \frac{0}{2}$$
$$x^2 - \frac{3}{2}x + 2 = 0$$

Isolate the variable terms.

$$x^2 - \frac{3}{2}x = -2$$
$$\left(-\frac{3}{2}\right)\left(\frac{1}{2}\right) = -\frac{3}{4}$$
$$\left(-\frac{3}{4}\right)^2 = \frac{9}{16}$$

Complete the square.

$$x^2 - \frac{3}{2}x + \frac{9}{16} = \frac{9}{16} - 2$$
$$\left(x - \frac{3}{4}\right)^2 = \frac{9}{16} - \frac{32}{16}$$
$$\left(x - \frac{3}{4}\right)^2 = -\frac{23}{16}$$

Use the square root property.

$$x - \tfrac{3}{4} = \pm \sqrt{-\tfrac{23}{16}}$$

$$x - \tfrac{3}{4} = \frac{\pm i \sqrt{23}}{4}$$

$$x = \tfrac{3}{4} \pm \frac{i \sqrt{23}}{4}$$

$$x = \frac{3 \pm i \sqrt{23}}{4}$$

Solving Quadratics by the Quadratic Formula

The following represents any quadratic equation:

$$ax^2 + bx + c = 0$$

This can be solved using the completing the square method to produce a formula that can then be applied to all quadratic equations.

Example 6: Solve $ax^2 + bx + c = 0$, $a \neq 0$, for x by using the completing the square method.

$$ax^2 + bx + c = 0$$

Get the coefficient of the squared term to be 1.

$$\frac{ax^2}{a} + \frac{bx}{a} + \frac{c}{a} = 0$$

Isolate the variable terms.

$$x^2 + \frac{b}{a}x + \frac{c}{a} = 0$$

$$x^2 + \frac{b}{a}x = -\frac{c}{a}$$

Complete the square.

$$\left(\frac{b}{a}\right)\left(\frac{1}{2}\right) = \frac{b}{2a}$$

$$\left(\frac{b}{2a}\right)^2 = \frac{b^2}{4a^2}$$

$$x^2 + \frac{b}{a}x + \frac{b^2}{4a^2} = \frac{b^2}{4a^2} - \frac{c}{a}$$

$$\left(x + \frac{b}{2a}\right)^2 = \frac{b^2}{4a^2} - \frac{4ac}{4a^2}$$

$$\left(x + \frac{b}{2a}\right)^2 = \frac{b^2 - 4ac}{4a^2}$$

Apply the square root property.

$$x + \frac{b}{2a} = \pm\sqrt{\frac{b^2 - 4ac}{4a^2}}$$

$$x + \frac{b}{2a} = \frac{\pm\sqrt{b^2 - 4ac}}{2a}$$

This last result is referred to as the **quadratic formula.** Remember, the quadratic formula can be used to solve all quadratic equations.

> *Quadratic Formula*
> $$x = \frac{-b \pm \sqrt{b^2 - 4ac}}{2a}$$

Example 7: Solve $2x^2 - 3x + 4 = 0$ by applying the quadratic formula.

$$2x^2 - 3x + 4 = 0$$

$$a = 2, \qquad b = -3, \qquad c = 4$$

$$x = \frac{-b \pm \sqrt{b^2 - 4ac}}{2a}$$

$$= \frac{-(-3) \pm \sqrt{(-3)^2 - 4(2)(4)}}{2(2)}$$

$$= \frac{3 \pm \sqrt{9 - 32}}{4}$$

$$= \frac{3 \pm \sqrt{-23}}{4}$$

$$= \frac{3 \pm i\sqrt{23}}{4}$$

Note that this is the same problem solved in Example 5 by completing the square. Here, however, it is solved by direct application of the quadratic formula.

Solving Equations in Quadratic Form

Any equation in the form $ax^2 + bx + c = 0$ is said to be in **quadratic form.**
This equation can then be solved by using the quadratic formula, by completing the square, or by factoring if it is factorable.

Example 8: Solve $x^4 - 13x^2 + 36 = 0$ by (a) factoring and (b) applying the quadratic formula.

(a)
$$x^4 - 13x^2 + 36 = 0$$
$$(x^2 - 4)(x^2 - 9) = 0$$
$$(x + 2)(x - 2)(x + 3)(x - 3) = 0$$

By the zero product rule,

$x + 2 = 0,$	$x - 2 = 0,$	$x + 3 = 0,$	$x - 3 = 0$
$x = -2$ or	$x = 2$ or	$x = -3$ or	$x = 3$

(b)
$$x^4 - 13x^2 + 36 = 0$$

is equivalent to $(x^2)^2 - 13(x^2) + 36 = 0$

$$a = 1, \qquad b = -13, \qquad c = 36$$

When applying the quadratic formula to equations in quadratic form,
you are solving for the variable name of the middle term. Thus, in
this case,

$$x^2 = \frac{-(-13) \pm \sqrt{(-13)^2 - 4(1)(36)}}{2(1)}$$

$$= \frac{13 \pm \sqrt{169 - 144}}{2}$$

$$= \frac{13 \pm \sqrt{25}}{2}$$

$$= \frac{13 \pm 5}{2}$$

$$x^2 = 9 \qquad \text{or} \qquad x^2 = 4$$

Using the square root property,

$$x = \pm \sqrt{9} \qquad \text{or} \qquad x = \pm \sqrt{4}$$
$$= \pm 3 \qquad\qquad = \pm 2$$

Example 9: Solve $x - 5\sqrt{x} - 6 = 0$ by (a) factoring and (b) applying the quadratic formula.

(a)
$$x - 5\sqrt{x} - 6 = 0$$
$$(\sqrt{x} - 6)(\sqrt{x} + 1) = 0$$
$$\sqrt{x} - 6 = 0 \quad \text{or} \quad \sqrt{x} + 1 = 0$$
$$\sqrt{x} = 6 \qquad\qquad \sqrt{x} = -1$$
$$(\sqrt{x})^2 = 6^2$$
$$x = 36$$

In the last step on the right, $\sqrt{x}$ must be a nonnegative value; therefore, $\sqrt{x} - 1$ has no solutions. The only solution is $x = 36$.

(b) $x - 5\sqrt{x} - 6 = 0$ is equivalent to $(\sqrt{x})^2 - 5(\sqrt{x}) - 6 = 0$

$\quad a = 1, \quad b = -5, \quad c = -6$

When applying the quadratic formula to this quadratic form equation, you are solving for $\sqrt{x}$.

$$\sqrt{x} = \frac{-(-5) \pm \sqrt{(-5)^2 - 4(1)(-6)}}{2(1)}$$
$$= \frac{5 \pm \sqrt{25 + 24}}{2}$$
$$= \frac{5 \pm \sqrt{49}}{2}$$
$$= \frac{5 \pm 7}{2}$$
$$\sqrt{x} = 6 \text{ or } \sqrt{x} = -1$$
$$x = 36$$

There is no solution for $\sqrt{x} = -1$. Thus, $x = 36$ is the only solution.

Solving Radical Equations

A **radical equation** is an equation in which a variable is under a radical. To solve a radical equation:

1. **Isolate the radical expression involving the variable. If more than one radical expression involves the variable, then isolate one of them.**

2. **Raise both sides of the equation to the index of the radical.**

3. **If there is still a radical equation, repeat Steps 1 and 2; otherwise, solve the resulting equation and *check* the answer in the original equation.**

By raising both sides of an equation to a power, some solutions may have been introduced that do not make the original equation true. These solutions are called **extraneous solutions.**

Example 10: Solve $\sqrt{3x^2 + 10x} - 5 = 0$.

$$\sqrt{3x^2 + 10x} - 5 = 0$$

Isolate the radical expression.

$$\sqrt{3x^2 + 10x} = 5$$

Raise both sides to the index of the radical; in this case, square both sides.

$$(\sqrt{3x^2 + 10x})^2 = (5)^2$$
$$3x^2 + 10x = 25$$
$$3x^2 + 10x - 25 = 0$$

This quadratic equation can now be solved either by factoring or by applying the quadratic formula.

Applying the quadratic formula,

$$3x^2 + 10x - 25 = 0$$

$$x = \frac{-10 \pm \sqrt{(10)^2 - 4(3)(-25)}}{2(3)}$$

$$= \frac{-10 \pm \sqrt{100 + 300}}{6}$$

$$= \frac{-10 \pm \sqrt{400}}{6}$$

$$= \frac{-10 \pm 20}{6}$$

$$x = \tfrac{5}{3} \text{ or } x = -5$$

Now, check the results.

If $x = \frac{5}{3}$,

$$\sqrt{3x^2 + 10x} - 5 = 0$$

$$\sqrt{3\left(\frac{5}{3}\right)^2 + 10\left(\frac{5}{3}\right)} - 5 \stackrel{?}{=} 0$$

$$\sqrt{3\left(\frac{25}{9}\right) + \frac{50}{3}} - 5 \stackrel{?}{=} 0$$

$$\sqrt{\frac{25}{3} + \frac{50}{3}} - 5 \stackrel{?}{=} 0$$

$$\sqrt{\frac{75}{3}} - 5 \stackrel{?}{=} 0$$

$$\sqrt{25} - 25 \stackrel{?}{=} 0$$

$$5 - 5 \stackrel{?}{=} 0$$

$$0 = 0 \checkmark$$

If $x = -5$,

$$\sqrt{3x^2 + 10x} - 5 = 0$$

$$\sqrt{3(-5)^2 + 10(-5)} - 5 \stackrel{?}{=} 0$$

$$\sqrt{3(25) - 50} - 5 \stackrel{?}{=} 0$$

$$\sqrt{75 - 50} - 5 \stackrel{?}{=} 0$$

$$\sqrt{25} - 5 \stackrel{?}{=} 0$$

$$5 - 5 \stackrel{?}{=} 0$$

$$0 = 0 \checkmark$$

The solution is $x = \frac{5}{3}$ or $x = -5$

Example 11: Solve $7 + \sqrt{a - 3} = 1$.

$$7 + \sqrt{a - 3} = 1$$

Isolate the radical expression.

$$\sqrt{a - 3} = -6$$

There is no solution, since $\sqrt{a - 3}$ cannot have a negative value.

Example 12: Solve $\sqrt{3x-5} + \sqrt{x-1} = 2$.

$$\sqrt{3x-5} + \sqrt{x-1} = 2$$

Isolate one of the radical expressions.

$$\sqrt{3x-5} = 2 - \sqrt{x-1}$$

Raise both sides to the index of the radical; in this case, square both sides.

$$(\sqrt{3x-5})^2 = (2 - \sqrt{x-1})^2$$
$$3x-5 = (2 - \sqrt{x-1})(2 - \sqrt{x-1})$$
$$= 4 - 2\sqrt{x-1} - 2\sqrt{x-1} + x - 1$$
$$= 3 - 4\sqrt{x-1} + x$$

This is still a radical equation. Isolate the radical expression.

$$3x-5 = 3 - 4\sqrt{x-1} + x$$
$$4\sqrt{x-1} = 8 - 2x$$

Raise both sides to the index of the radical; in this case, square both sides.

$$(4\sqrt{x-1})^2 = (8 - 2x)^2$$
$$16(x-1) = 64 - 32x + 4x^2$$
$$16x - 16 = 64 - 32x + 4x^2$$
$$0 = 4x^2 - 48x + 80$$

This can be solved either by factoring or by applying the quadratic formula.

Applying the quadratic formula,

$$4x^2 - 48x + 80 = 0$$

$$x = \frac{48 \pm \sqrt{(48)^2 - 4(4)(80)}}{2(4)}$$

$$= \frac{48 \pm \sqrt{2304 - 1280}}{8}$$

$$= \frac{48 \pm \sqrt{1024}}{8}$$

$$= \frac{48 \pm 32}{8}$$

$$x = 10 \qquad \text{or} \qquad x = 2$$

Check the solutions.

If $x = 10$,

$$\sqrt{3x - 5} + \sqrt{x - 1} = 2$$

$$\sqrt{3(10) - 5} + \sqrt{10 - 1} \overset{?}{=} 2$$

$$\sqrt{25} + \sqrt{9} \overset{?}{=} 2$$

$$5 + 3 \overset{?}{=} 2$$

$$8 = 2 \ \text{No}$$

So $x = 10$ is not a solution.

If $x = 2$,

$$\sqrt{3x - 5} + \sqrt{x - 1} = 2$$

$$\sqrt{3(2) - 5} + \sqrt{2 - 1} \overset{?}{=} 2$$

$$\sqrt{1} + \sqrt{1} \overset{?}{=} 2$$

$$1 + 1 \overset{?}{=} 2$$

$$2 = 2 \ \checkmark$$

The only solution is $x = 2$.

Example 13: Solve $\sqrt[3]{2x + 3} + 5 = 2$.

$$\sqrt[3]{2x + 3} + 5 = 2$$

Isolate the radical involving the variable.

$$\sqrt[3]{2x + 3} = -3$$

Since radicals with odd indexes can have negative answers, this problem does have solutions. Raise both sides of the equation to the index of the radical; in this case, cube both sides.

$$(\sqrt[3]{2x + 3})^3 = (-3)^3$$

$$2x + 3 = -27$$

$$2x = -30$$

$$x = -15$$

The check of the solution $x = -15$ is left to you.

Solving Quadratic Inequalities

To solve a quadratic inequality, follow these steps:

1. **Solve the inequality as though it were an equation.**

   The real solutions to the equation become boundary points for the solution to the inequality.

2. **Make the boundary points solid circles if the original inequality includes equality; otherwise, make the boundary points open circles.**

3. **Select points from each of the regions created by the boundary points. Replace these "test points" in the original inequality.**

4. **If a test point satisfies the original inequality, then the region that contains that test point is part of the solutions.**

5. **Represent the solution in graphic form and in solution set form.**

Example 14: Solve $(x - 3)(x + 2) > 0$.

Solve $(x - 3)(x + 2) = 0$. By the zero product property,

$$x - 3 = 0 \qquad \text{or} \qquad x + 2 = 0$$
$$x = 3 \qquad\qquad\qquad x = -2$$

Make the boundary points. Here, the boundary points are open circles because the original inequality does not include equality (see Figure 11-1).

Figure 11-1 Boundary points.

Select points from the different regions created (see Figure 11-2).

Figure 11-2 Three regions are created.

Try $x = -3$ $\qquad$ $x = 0$ $\qquad$ $x = 4$

See if the test points satisfy the original inequality.

$$(x-3)(x+2) > 0 \qquad (x-3)(x+2) > 0 \qquad (x-3)(x+2) > 0$$
$$(-3-3)(-3+2) \overset{?}{>} 0 \qquad (0-3)(0+2) \overset{?}{>} 0 \qquad (4-3)(-3+2) \overset{?}{>} 0$$
$$6 > 0 \checkmark \qquad\qquad -6 > 0 \text{ No} \qquad\qquad 6 > 0 \checkmark$$

Since $x = -3$ satisfies the original inequality, the region $x < -2$ is part of the solution. Since $x = 0$ does *not* satisfy the original inequality, the region $-2 < x < 3$ is *not* part of the solution. Since $x = 4$ satisfies the original inequality, the region $x > 3$ is part of the solution.

Represent the solution in graphic form and in solution set form. The graphic form is shown in Figure 11-3.

Figure 11-3   Solution to Example 14.

The solution set form is $\{x \mid x < -2 \text{ or } x > 3\}$

Example 15: Solve $9x^2 - 2 \le -3x$.

Solve $\qquad\qquad 9x^2 - 2 = -3x$
$$9x^2 + 3x - 2 = 0$$

By factoring,

$$(3x - 1)(3x + 2) = 0$$

$$3x - 1 = 0 \qquad \text{or} \qquad 3x + 2 = 0$$

$$x = \tfrac{1}{3} \qquad\qquad\qquad x = -\tfrac{2}{3}$$

Mark the boundary points using solid circles, as shown in Figure 11-4, since the original inequality includes equality.

Figure 11-4   Solid dots mean inclusion.

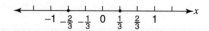

Select points from the regions created (see Figure 11-5).

Figure 11-5   Regions to test for Example 15.

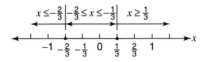

Try $x = -1$ $x = 0$ $x = 1$

See if the test points satisfy the original inequality.

$$9x^2 - 2 \le -3x \qquad\qquad 9x^2 - 2 \le -3x \qquad\qquad 9x^2 - 2 \le -3x$$

$$9(-1)^2 - 2 \overset{?}{\le} -3(-1) \quad 9(0)^2 - 2 \overset{?}{\le} -3(0) \quad 9(1)^2 - 2 \overset{?}{\le} -3(1)$$

$$9 - 2 \overset{?}{\le} 3 \qquad\qquad 0 - 2 \overset{?}{\le} 0 \qquad\qquad 9 - 2 \overset{?}{\le} -3$$

$$7 \le 3 \text{ No} \qquad\qquad -2 \le 0 \checkmark \qquad\qquad 7 \le -3 \text{ No}$$

Since $x = -1$ does *not* satisfy the original inequality, the region $x \le -\frac{2}{3}$ is *not* part of the solution. Since $x = 0$ does satisfy the original inequality, the region $-\frac{2}{3} \le x \le \frac{1}{3}$ is part of the solution. Since $x = 1$ does *not* satisfy the original inequality, the region $x \ge \frac{1}{3}$ is *not* part of the solution.

Represent the solution in graphic form and in solution set form. The graphic form is shown in Figure 11-6.

Figure 11-6   Solution to Example 15.

The set form is $\{x | -\frac{2}{3} \le x \le \frac{1}{3}\}$

Example 16: Solve $4t^2 - 9 < -4t$.

Solve $\qquad\qquad 4t^2 - 9 = -4t.$

$$4t^2 + 4t - 9 = 0$$

Since this quadratic is not easily factorable, the quadratic formula is used to solve it.

$$t = \frac{-4 \pm \sqrt{(4)^2 - 4(4)(-9)}}{2(4)}$$

$$= \frac{-4 \pm \sqrt{16 + 144}}{8}$$

$$= \frac{-4 \pm \sqrt{160}}{8}$$

$$= \frac{-4 \pm 4\sqrt{10}}{8}$$

Reduce by dividing out the common factor of 4.

$$t = \frac{-1 \pm \sqrt{10}}{2}$$

$$t = \frac{-1 + \sqrt{10}}{2} \quad \text{or} \quad t = \frac{-1 - \sqrt{10}}{2}$$

Since $\sqrt{10}$ is approximately 3.2,

$$t \approx \frac{-1 + 3.2}{2} \qquad\qquad t \approx \frac{-1 - 3.2}{2}$$
$$\text{or}$$
$$\approx 1.1 \qquad\qquad\qquad \approx -2.1$$

Mark the boundary points using open circles, as shown in Figure 11-7, since the original inequality does not include equality.

Figure 11-7 Open dots mean exclusion.

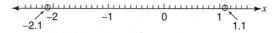

Select points from the different regions created (see Figure 11-8).

Figure 11-8 Regions to test for Example 16.

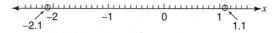

Try $t = -3$ $t = 0$ $t = 2$

See if the test points satisfy the original inequality.

$$4t^2 - 9 < -4t \qquad\qquad 4t^2 - 9 < -4t \qquad\qquad 4t^2 - 9 < -4t$$

$$4(-3)^2 - 9 \overset{?}{<} -4(-3) \qquad 4(0)^2 - 9 \overset{?}{<} -4(0) \qquad 4(2)^2 - 9 \overset{?}{<} -4(2)$$

$$27 < 12 \text{ No} \qquad\qquad -9 < 0 \checkmark \qquad\qquad 7 < -8 \text{ No}$$

Since $t = 3$ does *not* satisfy the original inequality, the region $t < \frac{(-1 - \sqrt{10})}{2}$ is *not* part of the solution. Since $t = 0$ does satisfy the original inequality, the region $\frac{(-1 - \sqrt{10})}{2} < t < \frac{(-1 + \sqrt{10})}{2}$ is part of the solution. Since $t = 2$ does *not* satisfy the original inequality, the region $t > \frac{(-1 + \sqrt{10})}{2}$ is *not* part of the solution.

Represent the solution in graphic form and in solution set form. The graphic form is shown in Figure 11-9.

Figure 11-9 Solution to Example 16.

The solution set form is $\left\{ t \, \middle| \, \dfrac{-1 - \sqrt{10}}{2} < t < \dfrac{-1 + \sqrt{10}}{2} \right\}$

Example 17: Solve $x^2 + 2x + 5 < 0$.

$$x^2 + 2x + 5 = 0$$

Since this quadratic is not factorable using rational numbers, the quadratic formula will be used to solve it.

$$x = \frac{-2 \pm \sqrt{(2)^2 - 4(1)(5)}}{2(1)}$$

$$= \frac{-2 \pm \sqrt{4 - 20}}{2}$$

$$= \frac{-2 \pm \sqrt{-16}}{2}$$

$$= \frac{-2 \pm 4i}{2}$$

$$= -1 \pm 2i$$

These are imaginary answers and cannot be graphed on a real number line. Therefore, the inequality $x^2 + 2x + 5 < 0$ has no real solutions.

Chapter Checkout

1. Solve by factoring:

$$3x^2 - 10x + 8 = 0$$

2. Solve by completing the square:

$$x^2 - 6x - 11 = 0$$

3. Solve by using the quadratic formula:

$$2x^2 - 3x - 8 = 0$$

Answers: 1. $2, \frac{4}{3}$ 2. $3 \pm 2\sqrt{5}$ 3. $\dfrac{3 \pm \sqrt{73}}{4}$

Chapter 12

CONIC SECTIONS

Chapter Check-In

- ❑ Analyzing circular equations
- ❑ Determining properties of parabolas
- ❑ Analyzing elliptical equations
- ❑ Finding axes of symmetry
- ❑ Determining asymptotes of hyperbolas

Conic sections are formed on a plane when that plane slices through a pair of cones stacked tip to tip. Whether the result is a circle, ellipse, parabola, or hyperbola depends only upon the angle at which the plane slices through. Conic sections are described mathematically by quadratic equations—some of which contain more than one variable.

The Four Conic Sections

When a double c-napped cone is sliced by a plane, the cross section formed by the plane and cone is called a **conic section.** The four main conic sections are the circle, the parabola, the ellipse, and the hyperbola (see Figure 12-1).

Figure 12-1   Creating conic sections.

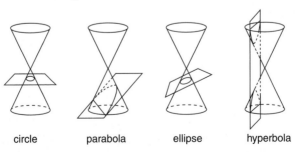

circle   parabola   ellipse   hyperbola

Circle

A **circle** is the set of points in a plane that are equidistant from one point. That one point is called the **center** of the circle, and the distance from it to any point on the circle is called the **radius** of the circle. The standard form for the equation of a circle with its center at (0, 0) and with a radius of length r is represented by the equation

$$x^2 + y^2 = r^2$$

Example 1: Graph $x^2 + y^2 = 16$.

Recognize that $x^2 + y^2 = 16$ is the equation of a circle centered at (0, 0) with $r^2 = 16$. So $r = 4$, as shown in Figure 12-2.

Figure 12-2   Circle in standard position.

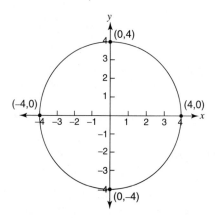

Example 2: Find the standard form for the equation of the circle centered (0, 0) with a radius of $\sqrt{5}$.

The standard form for a circle centered at (0, 0) with a radius of r is

$$x^2 + y^2 = r^2$$

Replacing r with $\sqrt{5}$, the equation becomes

$$x^2 + y^2 = (\sqrt{5})^2$$

$$= 5$$

Therefore, $x^2 + y^2 = 5$ is the standard form of the equation of a circle centered at (0, 0) with a radius of $\sqrt{5}$.

The standard form for a circle centered at (h, k) with a radius of r is

$$(x - h)^2 + (y - k)^2 = r^2$$

Example 3: Graph the equation $(x - 3)^2 + (y + 2)^2 = 25$.

This equation represents a circle centered at $(3, -2)$ with a radius of $\sqrt{25} = 5$, as shown in Figure 12-3.

Figure 12-3   Circle offset down and right.

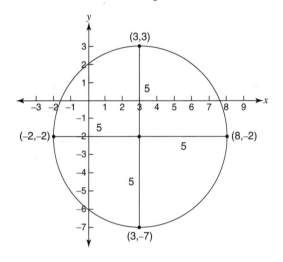

Example 4: Find the standard form for the equation of the circle centered at $(-6, 2)$ with a radius of $3\sqrt{2}$.

The standard form for the equation of a circle centered at (h, k) with radius r is

$$(x - h)^2 + (y - k)^2 = r^2$$

Replacing h with -6, k with 2, and r with $3\sqrt{2}$, the equation becomes

$$[x - (-6)]^2 + (y - 2)^2 = (3\sqrt{2})^2$$

Therefore, $(x + 6)^2 + (y - 2)^2 = 18$, which is the standard form of the equation of the circle centered at $(-6, 2)$ with radius $3\sqrt{2}$.

Example 5: From the equation given, find the center and radius for the following circle. Then graph the circle.

$$x^2 + y^2 - 8x + 12y - 12 = 0$$

This equation can be rewritten as

$$x^2 - 8x + y^2 + 12y = 12$$

Now, complete the square for each variable and add that amount to each side of the equation.

$$x^2 - 8x + \underline{16} + y^2 + 12y + \underline{36} = 12 + \underline{16} + \underline{36}$$
$$(x - 4)^2 + (y + 6)^2 = 64$$

This circle is centered at (4, –6) with a radius of 8, as shown in Figure 12-4.

Figure 12-4 The graph of Example 5.

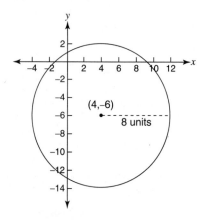

Parabola

A **parabola** is the set of points in a plane that are the same distance from a given point and a given line in that plane. The given point is called the **focus,** and the line is called the **directrix.** The midpoint of the perpendicular segment from the focus to the directrix is called the **vertex** of the parabola. The line that passes through the vertex and focus is called the **axis of symmetry** (see Figure 12-5.)

Figure 12-5 Two possible parabolas.

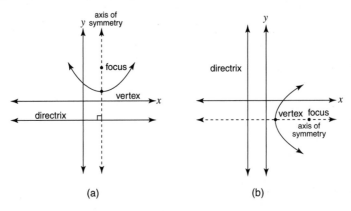

(a)   (b)

The equation of a parabola can be written in two basic forms

- **Form 1:** $y = a(x - h)^2 + k$
- **Form 2:** $x = a(y - k)^2 + h$

In Form 1, the parabola opens vertically. (It opens in the "y" direction.) If $a > 0$, it opens upward. Refer to Figure 12-5 (a). If $a < 0$, it opens downward. The distance from the vertex to the focus and from the vertex to the directrix line are the same. This distance is

$$\left| \frac{1}{4a} \right|$$

A parabola with its vertex at (h, k), opening vertically, will have the following properties.

- The focus will be at $\left(h, k + \frac{1}{4a} \right)$.
- The directrix will have the equation $y = k - \frac{1}{4a}$.
- The axis of symmetry will have the equation $x = h$.
- Its form will be $y = a(x - h)^2 + k$.

In Form 2, the parabola opens horizontally. (It opens in the "x" direction.) If $a > 0$, it opens to the right. Refer to Figure 12-5 (b). If $a < 0$, it opens to the left.

A parabola with its vertex at (h, k), opening horizontally, will have the following properties.

- The focus will be at $\left(h + \dfrac{1}{4a}, k\right)$.
- The directrix will have the equation $x = h - \dfrac{1}{4a}$.
- The axis of symmetry will have the equation $y = k$.
- Its form will be $x = a(y - k)^2 + h$.

Example 6: Draw the graph of $y = x^2$. State which direction the parabola opens and determine its vertex, focus, directrix, and axis of symmetry.

The equation $y = x^2$ can be written as

$$y = 1(x - 0)^2 + 0$$

so $a = 1$, $h = 0$, and $k = 0$. Since $a > 0$ and the parabola opens vertically, its direction is up (see Figure 12-6).

Vertex: $\qquad (h, k) = (0, 0)$

Focus: $\left(h, k + \dfrac{1}{4a}\right) = \left(0, 0 + \dfrac{1}{4(1)}\right)$

$\qquad\qquad\qquad = \left(0, \dfrac{1}{4}\right)$

Directrix: $\qquad y = k - \dfrac{1}{4a}$

$\qquad\qquad\qquad = 0 - \dfrac{1}{4}$

$\qquad\qquad\qquad = -\dfrac{1}{4}$

Axis of symmetry: $x = h$

$\qquad\qquad\qquad x = 0$

Figure 12-6 Properties of parabolas.

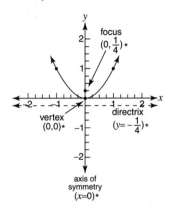

Example 7: Graph $x = -\frac{1}{8}(y + 2)^2 - 3$. State which direction the parabola opens and determine its vertex, focus, directrix, and axis of symmetry.

The equation $\qquad x = -\frac{1}{8}(y + 2)^2 - 3$

is the same as $\qquad x = -\frac{1}{8}[y - (-2)]^2 + (-3)$

$a = -\frac{1}{8}, \qquad h = -3, \quad k = -2$

Since $a < 0$ and the parabola opens horizontally, this parabola opens to the left (see Figure 12-7).

Vertex: $\qquad (h, k) = (-3, 2)$

Focus: $\left(h + \dfrac{1}{4a}, k\right) = \left(-3 + \dfrac{1}{4\left(-\frac{1}{8}\right)}, -2\right)$

$$= [-3 + (-2), -2]$$

$$= (-5, -2)$$

Directrix: $\qquad x = h - \dfrac{1}{4a}$

$$x = -3 - \dfrac{1}{4\left(-\frac{1}{8}\right)}$$

$$x = -3 - (-2)$$

$$x = -1$$

Axis of symmetry $y = k$

$$y = -2$$

Figure 12-7   The graph of Example 7.

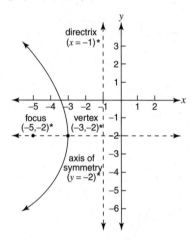

Example 8: Put the equation $x = 5y^2 - 30y + 11$ into the form

$$x = a(y - k)^2 + h$$

Determine the direction of opening, vertex, focus, directrix, and axis of symmetry.

$$x = 5y^2 - 30y + 11$$

Factor out the coefficient of y^2 from the terms involving y so that you can complete the square.

$$x = 5(y^2 - 6y) + 11$$

Completing the square within the parentheses adds $5(9) = 45$ to the right side. Add this amount to the left side to keep the equation balanced.

$$x + 45 = 5(y^2 - 6y + 9) + 11$$
$$x + 45 = 5(y - 3)^2 + 11$$

Subtract 45 from both sides.

$$x = 5(y - 3)^2 - 34$$

Direction: opens to the right ($a > 0$, opens horizontally)

Vertex: $\qquad (h, k) = (-34, 3)$

Focus: $\qquad \left(h + \frac{1}{4a}, k\right) = \left(-34 + \frac{1}{20}, 3\right)$

$\qquad\qquad\qquad = (-33\frac{19}{20}, 3)$

Directrix: $\qquad x = h - \frac{1}{4a}$

$\qquad\qquad x = -34 - \frac{1}{20}$

$\qquad\qquad x = -34\frac{1}{20}$

Axis of symmetry: $\qquad y = k$

$\qquad\qquad y = 3$

Ellipse

An **ellipse** is the set of points in a plane such that the sum of the distances from two fixed points in that plane stays constant. The two points are each called a **focus point.** The plural of focus is **foci.** The midpoint of the segment joining the foci is called the **center** of the ellipse. An ellipse has two axes of symmetry. The longer one is called the **major axis,** and the shorter one is called the **minor axis.** The two axes intersect at the center of the ellipse (see Figure 12-8).

Figure 12-8   Axes and foci of ellipses.

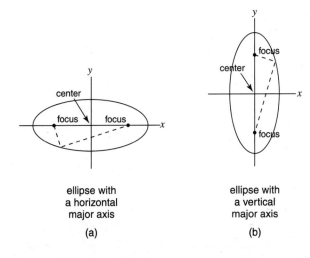

ellipse with
a horizontal
major axis

(a)

ellipse with
a vertical
major axis

(b)

The equation of an ellipse that is centered at (0, 0) and has its major axis along the *x*-axis has the following standard form.

$$\frac{x^2}{a^2} + \frac{y^2}{b^2} = 1 \qquad (a^2 > b^2)$$

The length of the major axis is $2|a|$, and the length of the minor axis is $2|b|$. The endpoints of the major axis are $(a, 0)$ and $(-a, 0)$ and are referred to as the **major intercepts.** The endpoints of the minor axis are $(0, b)$ and $(0, -b)$ and are referred to as the **minor intercepts.** If $(c, 0)$ and $(-c, 0)$ are the locations of the foci, then *c* can be found using the equation

$$c^2 = a^2 - b^2$$

If an ellipse has its major axis along the *y*-axis and is centered at (0, 0), the standard form becomes

$$\frac{x^2}{b^2} + \frac{y^2}{a^2} = 1 \qquad (a^2 > b^2)$$

The endpoints of the major axis become $(0, a)$ and $(0, -a)$. The endpoints of the minor axis become $(b, 0)$ and $(-b, 0)$. The foci are at $(0, c)$ and $(0, -c)$, with

$$c^2 = a^2 - b^2$$

When an ellipse is written in standard form, the major axis direction is determined by noting which variable has the larger denominator. The major axis either lies along that variable's axis or is parallel to that variable's axis.

Example 9: Graph the following ellipse. Find its major intercepts, length of the major axis, minor intercepts, length of the minor axis, and foci.

$$\frac{x^2}{4} + \frac{y^2}{9} = 1$$

This ellipse is centered at (0, 0). Since the larger denominator is with the *y* variable, the major axis lies along the *y*-axis.

Since	$a^2 = 9,$	$|a| = 3$
Since	$b^2 = 4,$ $\quad |b| = 2$	
Major intercepts:	(0, 3), (0, –3)	
Length of major axis:	$2|a| = 6$	
Minor intercepts:	(2, 0), (–2, 0)	

Length of minor axis: $2\,|b| = 4$

$$c^2 = a^2 - b^2$$
$$= 9 - 4$$
$$= 5$$
$$|c| = \sqrt{5}$$

Foci: $(0, \sqrt{5}), (0, -\sqrt{5})$

The graph of this ellipse is shown in Figure 12-9.

Figure 12-9   The graph of Example 9.

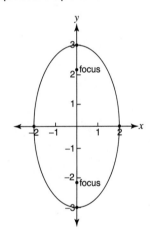

Example 10: Graph the following ellipse. Find its major and minor intercepts and its foci.

$$4x^2 + 25y^2 = 100$$

Write $4x^2 + 25y^2 = 100$ in standard form by dividing each side by 100.

$$\frac{4x^2}{100} + \frac{25y^2}{100} = \frac{100}{100}$$

$$\frac{x^2}{25} + \frac{y^2}{4} = 1$$

This ellipse is centered at (0, 0). Since the larger denominator is with the *x* variable, the major axis lies along the *x*-axis

$$a^2 = 25 \qquad b^2 = 4 \qquad c^2 = a^2 - b^2$$
$$|a| = 5 \qquad |b| = 2 \qquad \quad = 25 - 4$$
$$= 21$$
$$|c| = \sqrt{21}$$

Major intercepts: (5, 0), (−5, 0)

Minor intercepts: (0, 2), (0, −2)

Foci: ($\sqrt{21}$, 0), (−$\sqrt{21}$, 0)

The graph of this ellipse is shown in Figure 12-10.

Figure 12-10   The graph of Example 10.

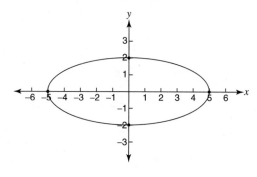

The standard form for an ellipse centered at (*h, k*) with its major axis parallel to the *x*-axis is

$$\frac{(x-h)^2}{a^2} + \frac{(y-k)^2}{b^2} = 1 \qquad (a^2 > b^2)$$

Major intercepts: (*h* + *a*,*k*), (*h* − *a*,*k*)

Minor intercepts: (*h*,*k* + *b*), (*h*,*k* −*b*)

Foci: (*h*+*c*,*k*), (*h* − *c*,*k*) with $c^2 = a^2 - b^2$

The standard form for an ellipse centered at (*h, k*) with a major axis parallel to the *y*-axis is

$$\frac{(x-h)^2}{a^2} + \frac{(y-k)^2}{b^2} = 1 \qquad (a^2 > b^2)$$

Major intercepts:	$(h, k + a)$	$(h, k - a)$	
Minor intercepts:	$(h + b, k)$	$(h - b, k)$	
Foci:	$(h + c, k)$	$(h - c, k)$	with $c^2 = a^2 - b^2$

Example 11: Graph the following ellipse. Find its center, major and minor intercepts, and foci.

$$\frac{(x-2)^2}{36} + \frac{(y+1)^2}{25} = 1$$

Center: $(2, -1)$

$$a^2 = 36 \qquad b^2 = 25 \qquad c^2 = a^2 - b^2$$
$$|a| = 6 \qquad |b| = 5 \qquad = 36 - 25$$
$$= 11$$
$$|c| = \sqrt{11}$$

Major intercepts:
$$(2 + 6, -1) = (8, -1)$$
$$(2 - 6, -1) = (-4, -1)$$

Minor intercepts:
$$(2, -1 + 5) = (2, 4)$$
$$(2, -1 - 5) = (2, -6)$$

Foci:
$$(2 + \sqrt{11}, -1), (2 - \sqrt{11}, -1)$$

The graph of this ellipse is shown in Figure 12-11.

Figure 12-11   The graph of Example 11.

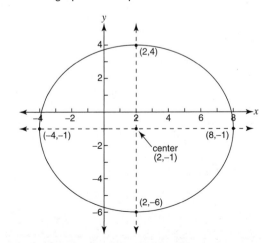

Example 12: An ellipse has the following equation.

$$16x^2 + 25y^2 + 32x - 150y = 159$$

Find the coordinates of its center, major and minor intercepts, and foci. Then graph the ellipse.

$$16x^2 + 25y^2 + 32x - 150y = 159$$

Rearrange terms.

$$16x^2 + 32x + 25y^2 - 150y = 159$$

Factor out the coefficient of each of the squared terms.

$$16(x^2 + 2x) + 25(y^2 - 6y) = 159$$

Complete the square within each set of parentheses and add the same amount to both sides of the equation.

$$16(x^2 + 2x + \underline{1}) + 25(y^2 - 6y + \underline{9}) = 159 + \underline{16(1)} + \underline{25(9)}$$
$$16(x + 1)^2 + 25(y - 3)^2 = 400$$

Divide each side by 400.

$$\frac{\overset{1}{\cancel{16}}(x + 1)^2}{\underset{25}{\cancel{400}}} + \frac{\overset{1}{\cancel{25}}(y - 3)^2}{\underset{16}{\cancel{400}}} = \frac{400}{400}$$

$$\frac{(x + 1)^2}{25} + \frac{(y - 3)^2}{16} = 1$$

Center $(-1, 3)$: Since the x variable has the larger denominator, the major axis is parallel to the x-axis.

$$a^2 = 25 \qquad b^2 = 16 \qquad c^2 = a^2 - b^2$$
$$|a| = 5 \qquad |b| = 4 \qquad\qquad = 25 - 16$$
$$= 9$$
$$|c| = 3$$

Major intercepts: $\quad (-1 + 5, 3) = (4, 3)$

$\qquad\qquad\qquad\quad (-1 - 5, 3) = (-6, 3)$

Minor intercepts: $\quad (-1, 3 + 4) = (-1, 7)$

$\qquad\qquad\qquad\quad (-1, 3 - 4) = (-1, -1)$

Foci: $\qquad\qquad\quad (-1 + 3, 3) = (2, 3)$

$\qquad\qquad\qquad\quad (-1 - 3, 3) = (-4, 3)$

The graph of this ellipse is shown in Figure 12-12.

Figure 12-12   The graph of Example 12.

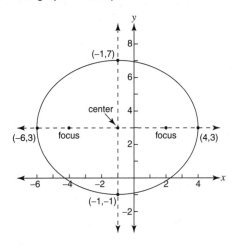

Hyperbola

A **hyperbola** is the set of all points in a plane such that the absolute value of the difference of the distances between two fixed points stays constant. The two given points are the **foci** of the hyperbola, and the midpoint of the segment joining the foci is the **center** of the hyperbola. The hyperbola looks like two opposing "U-shaped" curves, as shown in Figure 12-13.

Figure 12-13   Properties of hyperbolas.

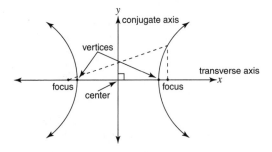

A hyperbola has two axes of symmetry (refer to Figure 12-13). The axis along the direction the hyperbola opens is called the **transverse axis.** The **conjugate axis** passes through the center of the hyperbola and is perpendicular to the transverse axis. The points of intersection of the hyperbola and the transverse axis are called the **vertices** (singular, **vertex**) of the hyperbola.

A hyperbola centered at (0, 0) whose transverse axis is along the *x*-axis has the following equation as its standard form.

$$\frac{x^2}{a^2} - \frac{y^2}{b^2} = 1$$

where (*a*, 0) and (−*a*, 0) are the vertices and (*c*, 0) and (−*c*, 0) are its foci. In the hyperbola, $c^2 = a^2 + b^2$.

As points on a hyperbola get farther from its center, they get closer and closer to two lines called **asymptote lines.** The asymptote lines are used as guidelines in sketching the graph of a hyperbola. To graph the asymptote lines, form a rectangle by using the points (−*a*, *b*), (−*a*, −*b*), (*a*, *b*), and (*a*, −*b*) and draw its diagonals as extended lines.

For the hyperbola centered at (0, 0) whose transverse axis is along the *x*-axis, the equation of the asymptote lines becomes

$$y = \pm \frac{b}{a} x$$

Example 13: Graph the following hyperbola. Find its center, vertices, foci, and the equations of its asymptote lines.

$$\frac{x^2}{16} - \frac{y^2}{25} = 1$$

This is a hyperbola with center at (0, 0), and its transverse axis is along the *x*-axis.

$a^2 = 16$	$b^2 = 25$	$c^2 = a^2 + b^2$
$|a| = 4$	$|b| = 5$	$= 16 + 25$
		$= 41$
		$|c| = \sqrt{41}$

Vertices: (−4, 0) (4, 0)

Foci: (−$\sqrt{41}$, 0) ($\sqrt{41}$, 0)

Equations of asymptote lines: $y = \pm \frac{5}{4} x$

The graph of this hyperbola is shown in Figure 12-14.

Figure 12-14   Center, asymptotes, foci, and vertices.

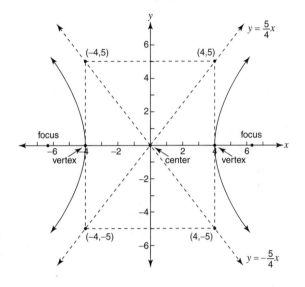

A hyperbola centered at (0, 0) whose transverse axis is along the y-axis has the following equation as its standard form.

$$\frac{y^2}{a^2} - \frac{x^2}{b^2} = 1$$

The vertices are now (0, a) and (0, $-a$). The foci are at (0, c) and (0, $-c$) with $c^2 = a^2 + b^2$. The asymptote lines have equations

$$y = \pm \frac{a}{b} x$$

In general, when a hyperbola is written in standard form, the transverse axis is along, or parallel to, the axis of the variable that is *not* being subtracted.

Example 14: Graph the following hyperbola and find its center, vertices, foci, and equations of the asymptote lines.

$$\frac{y^2}{4} - \frac{x^2}{9} = 1$$

This is a hyperbola with its center at (0, 0) and its transverse axis along the *y*-axis, since the *y* variable is not being subtracted.

$$a^2 = 4 \qquad b^2 = 9 \qquad c^2 = a^2 + b^2$$
$$|a| = 2 \qquad |b| = 3 \qquad = 4 + 9$$
$$ = 13$$
$$ |c| = \sqrt{13}$$

Vertices: $\qquad\qquad\qquad$ (0, 2), $\qquad$ (0, –2)

Foci: $\qquad\qquad\qquad$ $(0, \sqrt{13}),\quad (0, -\sqrt{13})$

Equations of asymptote lines: $\quad y = \pm\frac{2}{3}x$

The graph of this hyperbola is shown in Figure 12-15.

Figure 12-15 The graph of Example 14.

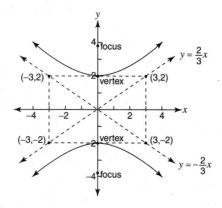

Hyperbola centered at (*h*, *k*) will have the following standard equations:

■ If the transverse axis is horizontal, then

$$\frac{(x - h)^2}{a^2} - \frac{(y - k)^2}{b^2} = 1$$

In this case, the vertices are at (*h* + *a*, *k*) and (*h* – *a*, *k*). The foci are at (*h* + *c*, *k*) and (*h* – *c*, *k*) where $c^2 = a^2 + b^2$.

■ If the transverse axis is vertical,

$$\frac{\left(y-k\right)^2}{a^2} - \frac{\left(x-h\right)^2}{b^2} = 1$$

The vertices are at $(h, k + a)$ and $(h, k - a)$, and the foci are at $(h, k + c)$ and $(h, k - c)$ where $c^2 = a^2 + b^2$.

Example 15: Graph $\frac{(x-2)^2}{16} - \frac{\left(y+3\right)^2}{25} = 1$.

Center: $(2, -3)$

The transverse axis is horizontal.

Vertices: $(2 + 4, -3) = (6, -3)$

$(2 - 4, -3) = (-2, -3)$

Find the coordinates of the corners of the rectangle to help make the asymptote lines.

$$(2 + 4, -3 + 5) = (6, 2)$$
$$(2 - 4, -3 + 5) = (-2, 2)$$
$$(2 - 4, -3 - 5) = (6, -8)$$
$$(2 - 4, -3 - 5) = (-2, -8)$$

The graph of this hyperbola is shown in Figure 12-16.

Figure 12-16   The graph of Example 15.

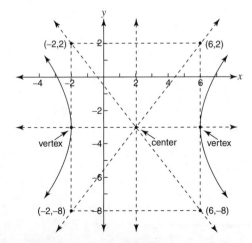

The equation $xy = 16$ also represents a hyperbola. This hyperbola has its center at $(0, 0)$, and its transverse axis is the line $y = x$. The asymptotes are the x- and y-axes. Its vertices are at $(\sqrt{16}, \sqrt{16}) = (4, 4)$ and $(-\sqrt{16}, -\sqrt{16}) = (-4, -4)$. The graph of this hyperbola is shown in Figure 12-17.

Figure 12-17   Hyperbola in quadrants I and III.

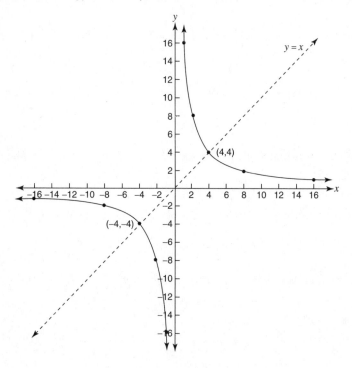

Example 16: Graph $xy = -4$.

This hyperbola has its center at $(0, 0)$. The transverse axis is the line $y = -x$. The asymptotes are the x- and y-axes. The vertices are at $(\sqrt{4}, -\sqrt{4}) = (2, -2)$ and $(-\sqrt{4}, \sqrt{4}) = (-2, 2)$. The graph of this hyperbola is shown in Figure 12-18.

Figure 12-18   Hyperbola in quadrants II and IV.

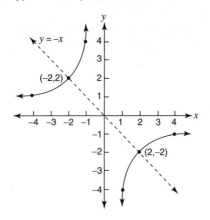

Chapter Checkout

1. Write the standard form for the circle with center at (5,3) and radius $2\sqrt{5}$.

2. Examine the following equation:

$$x^2 - x - 6 = 0$$

    a. What shape is its graph?
    b. Where is the axis of symmetry?
    c. What are the coordinates of the vertex?

3. Following is the equation for an ellipse:

$$\frac{x^2}{16} + \frac{y^2}{36} = 1$$

    a. Is its major axis horizontal or vertical?
    b. What are the coordinates of its foci?
    c. What is the length of the major axis?
    d. What is the length of the minor axis?

Answers: 1. $(x-5)^2 + (y-3)^2 = 20$ 2. a. parabola b. $x = \frac{1}{2}$, c. $\left(\frac{1}{2} - 6\frac{1}{4}\right)$
3. a. vertical b. $\left(0, 2\sqrt{5}\right)$ c. $\left(0, -2\sqrt{5}\right)$ d. 12, 8

Chapter 13
QUADRATIC SYSTEMS

Chapter Check-In

❑ Solving systems algebraically

❑ Solving systems graphically

❑ Solving systems of inequalities

Earlier chapters explain systems of linear equations and explore various strategies for solving them. This chapter discusses quadratic systems, and the options for solution have narrowed. Don't get the idea that quadratic systems cannot be solved by substitution the way linear ones can. It's just that if both equations are second degree, the substitution method becomes far too convoluted and time-consuming to be worthwhile. On the other hand, if you are dealing with a system containing a quadratic and a linear equation, substitution may be the method of choice. Don't lose sight of quadratic inequalities either. This chapter also deals with them.

Systems of Equations Solved Algebraically

When given two equations in two variables, there are essentially two algebraic methods for solving them. One is substitution, and the other is elimination.

Example 1: Solve the following system of equations algebraically.

$$(1)\ x^2 + 2y^2 = 10$$
$$(2)\ 3x^2 - y^2 = 9$$

This system is more easily solved using the elimination method.

equation (1)	$x^2 + 2y^2 = 10$
2 times equation (2)	$\underline{6x^2 - 2y^2 = 18}$
Add the results	$7x^2\qquad = 28$
	$x^2 = 4$
	$x = \pm 2$

Using equation (1),

$$\text{If } x = 2 \qquad\qquad\qquad \text{If } x = -2$$

$$x^2 + 2y^2 = 10 \qquad\qquad x^2 + 2y^2 = 10$$

$$(2)^2 + 2y^2 = 10 \qquad\qquad (-2)^2 + 2y^2 = 10$$

$$4 + 2y^2 = 10 \qquad\qquad 4 + 2y^2 = 10$$

$$2y^2 = 6 \qquad\qquad\qquad 2y^2 = 6$$

$$y^2 = 3 \qquad\qquad\qquad y^2 = 3$$

$$y = \pm\sqrt{3} \qquad\qquad\qquad y = \pm\sqrt{3}$$

$$(2, \sqrt{3})\ (2, -\sqrt{3}) \qquad\quad (-2, \sqrt{3})\ (-2, -\sqrt{3})$$

The solution consists of the above four ordered pairs. If these equations were graphed, these ordered pairs would represent the points of intersection of the graphs.

Example 2: Solve the following system of equations algebraically.

$$(1) \quad x^2 + y^2 = 100$$

$$(2) \quad x - y = 2$$

This system is more easily solved using substitution. Solve equation (2) for x; then substitute that result for x in equation (1).

Solving equation (2) for x,

$$x - y = 2$$

$$x = y + 2$$

Substituting into equation (1),

$$x^2 + y^2 = 100$$

$$(y + 2)^2 + y^2 = 100$$

$$y^2 + 4y + 4 + y^2 = 100$$

$$2y^2 + 4y + 4 = 100$$

$$2y^2 + 4y - 96 = 0$$

$$2(y^2 + 2y - 48) = 0$$

$$2(y - 6)(y + 8) = 0$$

$$y - 6 = 0 \qquad \text{or} \qquad y + 8 = 0$$
$$y = 6 \qquad\qquad\qquad y = -8$$

Using equation (2),

$$\text{If } y = 6 \qquad\qquad\qquad \text{If } y = -8$$
$$x - y = 2 \qquad\qquad\qquad x - y = 2$$
$$x - 6 = 2 \qquad\qquad\qquad x - (-8) = 2$$
$$x = 8 \qquad\qquad\qquad x = -6$$
$$(8, 6) \qquad\qquad\qquad (-6, -8)$$

The solution consists of the above two ordered pairs. If this system of equations were graphed, these two points would represent the points where the graphs would intersect.

Systems of Equations Solved Graphically

Graphs can be used to solve systems of equations. This method, however, usually allows only approximate solutions, whereas the algebraic method arrives at exact solutions.

Example 3: Solve the following system of equations graphically.

$$(1) \quad x^2 + 2y^2 = 10$$
$$(2) \quad 3x^2 - y^2 = 9$$

Equation (1) is the equation of an ellipse. Convert the equation into standard form.

$$x^2 + 2y^2 = 10$$
$$\frac{x^2}{10} + \frac{2y^2}{10} = \frac{10}{10}$$
$$\frac{x^2}{10} + \frac{y^2}{5} = 1$$

The major intercepts are at $(\sqrt{10}, 0)$ and $(-\sqrt{10}, 0)$, and the minor intercepts are at $(0, \sqrt{5})$ and $(0, -\sqrt{5})$.

Equation (2) is the equation of a hyperbola. Convert the equation into standard form.

$$3x^2 - y^2 = 9$$

$$\frac{3x^2}{9} - \frac{y^2}{9} = \frac{9}{9}$$

$$\frac{x^2}{3} - \frac{y^2}{9} = 1$$

The transverse axis is horizontal and the vertices are at $(\sqrt{3}, 0)$ and $(-\sqrt{3}, 0)$, as shown in Figure 13-1.

Figure 13-1   Approximate solutions to hyperbola and ellipse.

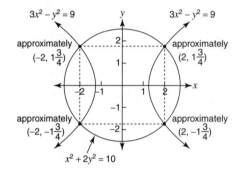

The approximate answers are

$$\left\{\left(-2, 1\tfrac{3}{4}\right), \left(-2, -1\tfrac{3}{4}\right), \left(2, 1\tfrac{3}{4}\right), \left(2, -1\tfrac{3}{4}\right)\right\}$$

The exact answers are

$$\left\{\left(-2, \sqrt{3}\right), \left(-2, -\sqrt{3}\right), \left(2, \sqrt{3}\right), \left(2, -\sqrt{3}\right)\right\}$$

Refer to Example 1 for the algebraic approach to this problem; it gives the exact answers.

Example 4: Solve the following system of equations graphically.

$$(1) \qquad x^2 + y^2 = 100$$
$$(2) \qquad x - y = 2$$

Equation (1) is the equation of a circle centered at (0, 0) with a radius of 10. Equation (2) is the equation of a line. The solutions are

$$\{(-6, -8), (8, 6)\}$$

The graph is shown in Figure 13-2.

Figure 13-2 Circle with intersecting line.

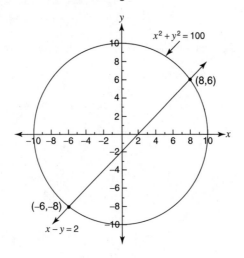

Refer to Example 2 for the algebraic approach to this problem.

Systems of Inequalities Solved Graphically

To graph the solutions of a system of inequalities, graph each inequality and find the intersections of the two graphs.

Example 5: Graph the solutions for the following system.

$$(1)\ x^2 + y^2 \leq 16$$
$$(2)\ y \leq x^2 + 2$$

Equation (1) is the equation of a circle centered at (0, 0) with a radius of 4. Graph the circle; then select a test point not on the circle and place it into the original inequality. If that result is true, then shade the region where the test point is located. Otherwise, shade the other region. Use (0, 0) as a test point.

$$x^2 + y^2 \leq 16$$
$$0^2 + 0^2 \leq 16$$

This is a true statement. Therefore, the interior of the circle is shaded. In Figure 13-3 (a), this shading is done with horizontal lines.

Equation (2) is the equation of a parabola opening upward with its vertex at (0, 2). Use (0, 0) as a test point.

$$y \leq x^2 + 2$$
$$0 \leq 0^2 + 2 \checkmark$$

This is a true statement. Therefore, shade the exterior of the parabola. In Figure 13-3 (a), this shading is done with vertical lines. The region with both shadings represents the solutions of the systems of inequalities. That solution is shown by the shading on the right side of Figure 13-3 (b).

Figure 13-3   Shading shows solutions.

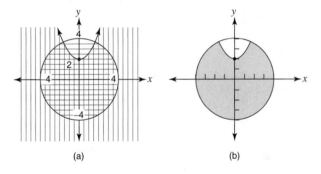

(a)   (b)

Example 6: Solve the following system of inequalities graphically.

$$(1) \qquad \frac{x^2}{36} + \frac{y^2}{25} \leq 1$$

$$(2) \qquad \frac{y^2}{4} - \frac{x^2}{1} > 1$$

Equation (1) is the equation of an ellipse centered at (0, 0) with major intercepts at (6, 0) and (–6, 0) and minor intercepts at (0, 5) and (0, –5). Use (0, 0) as a test point.

$$\frac{x^2}{36} + \frac{y^2}{25} \leq 1$$

$$\frac{0^2}{36} + \frac{0^2}{25} \leq 1 \checkmark$$

This is a true statement. Therefore, shade the interior of the ellipse. In Figure 13-4 (a), this shading is done horizontally.

Equation (2) is the equation of a hyperbola centered at (0, 0) opening vertically with vertices at (0, 2) and (0, –2). Use (0, 0) as a test point.

$$\frac{y^2}{4} - \frac{x^2}{1} > 1$$

$$\frac{0^2}{4} - \frac{0^2}{1} > 1 \ \text{No}$$

This is not a true statement. Therefore, shade the area inside the curves of the hyperbola. In Figure 13-4 (a), this shading is done vertically. The region with both shadings represents the solution to the system of inequalities. That solution is shown by the shading in Figure 13-4 (b).

Figure 13-4 Solution to Example 6.

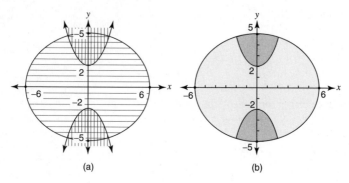

(a)   (b)

Chapter Checkout

1. Solve by substitution:

$$x^2 + 3y^2 = 172$$
$$2x - y = 3$$

2. Solve by elimination:

$$2x^2 - 3y^2 = 5$$
$$3x^2 + 2y^2 = 66$$

3. Solve the following system any way you can.

$$16x^2 + 4y^2 = 10$$
$$8y^2 - 32x^2 = -16$$

Answers: 1. $x = 5$, $y = 7$ 2. $x = 4$, $y = -3$ 3. $x = \frac{3}{4}$, $y = \frac{1}{2}$

Chapter 14

EXPONENTIAL AND LOGARITHMIC FUNCTIONS

Chapter Check-In

❑ Defining exponential functions

❑ Examining logarithmic functions

❑ Examining properties of logarithms

❑ Solving exponential and logarithmic equations

Exponential functions grow exponentially—that is, very, very quickly. Two squared is 4; 2 cubed is 8, but by the time you get to 2^7, you have, in four small steps from 8, already reached 128, and it only grows faster from there. Four more steps, for example, brings the value to 2,048.

Logarithmic functions are the mathematics of exponents. This chapter introduces the concept of logarithms and how any exponential function can be expressed in logarithmic form. Similarly, all logarithmic functions can be rewritten in exponential form. Logarithms are really useful in permitting us to work with very large numbers while manipulating numbers of a much more manageable size. While logarithms of the base 10 and the base e are those most often dealt with in mathematics, this chapter deals with logs to other bases as well.

Exponential Functions

Any function defined by $y = b^x$, where $b > 0$, $b \neq 1$, and x is a real number, is called an **exponential function.**

Example 1: Graph $y = 2^x$.

First find a sufficient number of ordered pairs to see the shape of the graph.

x	$2^x = y$	(x, y)
−3	$2^{-3} = \frac{1}{8}$	$(-3, \frac{1}{8})$
−2	$2^{-2} = \frac{1}{4}$	$(-2, \frac{1}{4})$
−1	$2^{-1} = \frac{1}{2}$	$(-1, \frac{1}{2})$
0	$2^0 = 1$	$(0, 1)$
1	$2^1 = 2$	$(1, 2)$
2	$2^2 = 4$	$(2, 4)$
3	$2^3 = 8$	$(3, 8)$

Find these points and connect them to form a smooth curve. There is no value for x that will make y become zero. The more negative x becomes, the smaller y becomes. The negative x-axis becomes an asymptote for this function. The graph is shown in Figure 14-1.

Figure 14-1 Exponential growth of 2.

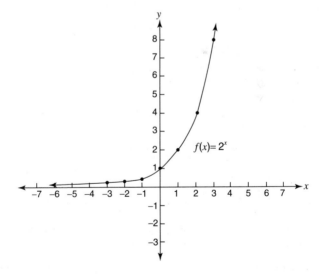

Example 2: Graph $f(x) = (\frac{1}{2})^x$.

Make a chart and graph the ordered pairs.

x	$\left(\frac{1}{2}\right)^x = f(x)$	$[x, f(x)]$
-3	$\left(\frac{1}{2}\right)^{-3} = 8$	$(-3, 8)$
-2	$\left(\frac{1}{2}\right)^{-2} = 4$	$(-2, 4)$
-1	$\left(\frac{1}{2}\right)^{-1} = 2$	$(-1, 2)$
0	$\left(\frac{1}{2}\right)^{0} = 1$	$(0, 1)$
1	$\left(\frac{1}{2}\right)^{1} = \frac{1}{2}$	$(1, \frac{1}{2})$
2	$\left(\frac{1}{2}\right)^{2} = \frac{1}{4}$	$(2, \frac{1}{4})$
3	$\left(\frac{1}{2}\right)^{3} = \frac{1}{8}$	$(3, \frac{1}{8})$

The graph is shown in Figure 14-2.

Figure 14-2   Exponential growth of $\frac{1}{2}$.

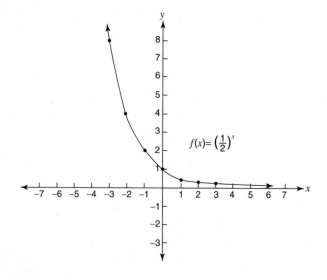

$$f(x) = \left(\frac{1}{2}\right)^x$$

All exponential functions, $f(x) = b^x$, $b > 0$, $b \neq 1$, will contain the ordered pair $(0, 1)$, since $b^0 = 1$ for all $b \neq 0$. Exponential functions with $b > 1$ will have a basic shape like that in the graph shown in Figure 14-1, and exponential functions with $b < 1$ will have a basic shape like that of Figure 14-2.

The graph of $x = b^y$ is called the **inverse** of the graph of $y = b^x$ because the x and y variables are interchanged. Remember, the graphs of inverses are symmetrical around the line $y = x$. That is, if the graph of $y = b^x$ is "folded over" the line $y = x$ and then retraced, it creates the graph of $x = b^y$. Whatever ordered pairs satisfy $y = b^x$, the reversed ordered pairs would satisfy $x = b^y$.

Example 3: Graph $y = 3^x$ and $x = 3^y$ on the same set of axes.

x	$3^x = y$	(x, y)	x	$3^y = x$	(x, y)
-3	$(3)^{-3} = \frac{1}{27}$	$(-3, \frac{1}{27})$	$\frac{1}{27}$	$(3)^{-3} = \frac{1}{27}$	$(\frac{1}{27}, -3)$
-2	$(3)^{-2} = \frac{1}{9}$	$(-2, \frac{1}{9})$	$\frac{1}{9}$	$(3)^{-2} = \frac{1}{9}$	$(\frac{1}{9}, -2)$
-1	$(3)^{-1} = \frac{1}{3}$	$(-1, \frac{1}{3})$	$\frac{1}{3}$	$(3)^{-1} = \frac{1}{3}$	$(\frac{1}{3}, -1)$
0	$(3)^0 = 1$	$(0, 1)$	1	$(3)^0 = 1$	$(1, 0)$
1	$(3)^1 = 3$	$(1, 3)$	3	$(3)^1 = 3$	$(3, 1)$
2	$(3)^2 = 9$	$(2, 9)$	9	$(3)^2 = 9$	$(9, 2)$
3	$(3)^3 = 27$	$(3, 27)$	27	$(3)^3 = 27$	$(27, 3)$

The graph of $x = 3^y$ is shown in Figure 14-3.

Figure 14-3 In this graph, $y = 3^x$ and $x = 3^y$.

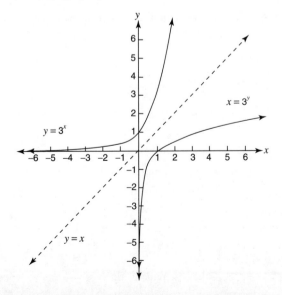

Logarithmic Functions

If $x = 2^y$ were to be solved for y, so that it could be written in function form, a new word or symbol would need to be introduced. If $x = 2^y$, then $y =$ (the power on base 2) to equal x. The word **logarithm,** abbreviated **log,** is introduced to satisfy this need.

$$y = \text{(the power on base 2) to equal } x$$

is rewritten as $y = \log_2 x$

This is read as "y equals the log of x, base 2" or "y equals the log, base 2, of x."

A **logarithmic function** is a function of the form

$$y = \log_b x \quad x > 0, \text{ where } b > 0 \text{ and } b \neq 1$$

which is read "y equals the log of x, base b" or "y equals the log, base b, of x."

$$y = \log_b x \quad \text{is equivalent to} \quad x = b^y$$

the base remains the base

In both forms, $x > 0$ and $b > 0$, $b \neq 1$. There are no restrictions on y.

Example 4: Rewrite each exponential equation in its equivalent logarithmic form. The solutions follow.

(a) $5^2 = 25$ (b) $4^{-3} = \frac{1}{64}$ (c) $\left(\frac{1}{2}\right)^{-4} = 16$

(a)
$$5^2 = 25 \quad \text{becomes} \quad 2 = \log_5 25$$
the base remains the base

(b)
$$4^{-3} = \frac{1}{64} \quad \text{becomes} \quad -3 = \log_4\left(\frac{1}{64}\right)$$
the base remains the base

(c)
$$\left(\frac{1}{2}\right)^{-4} = 16 \quad \text{becomes} \quad -4 = \log_{(1/2)} 16$$
the base remains the base

Example 5: Rewrite each logarithmic equation in its equivalent exponential form. The solutions follow.

$$\text{(a) } \log_6 36 = 2 \qquad \text{(b) } \log_a m = p$$

(a)
$$\log_6 36 \quad \text{becomes} \quad 6^2 = 36$$
$$\text{base}$$

(b)
$$\log_a m = p \quad \text{becomes} \quad a^p = m$$
$$\text{base}$$

Example 6: Solve the following equations, if possible.

$$\text{(a) } \log_7 49 = y^2 \qquad \text{(c) } \log_y 8 = 3 \qquad \text{(e) } \log_3(-9) = y$$
$$\text{(b) } \log_2(\tfrac{1}{8}) = y \qquad \text{(d) } \log_4 y = -2$$

(a) $\log_7 49 = y$ becomes $7^y = 49$

Since $49 = 7^2$, $7^y = 7^2$

So $y = 2$

(b) $\log_2(\tfrac{1}{8}) = y$ becomes $2^y = \tfrac{1}{8}$

Since $\tfrac{1}{8} = 2^{-3}$, $2^y = 2^{-3}$

So $y = -3$

(c) $\log_y 8 = 3$ becomes $y^3 = 8$

Since $8 = 2^3$ $y^3 = 2^3$

So $y = 2$

(d) $\log_4 y = -2$ becomes $4^{-2} = y$

Since $4^{-2} = \tfrac{1}{16}$, $\tfrac{1}{16} = y$

(e) $\log_3(-9) = y$ becomes $3^y = -9$

This is not possible, since 3^y will always be a positive result. Recall that logarithms have only a positive domain; therefore, –9 is not in the domain of a logarithm.

The bases used most often when working with logarithms are base 10 and base e. (The letter e represents an irrational number that has many applications in mathematics and science. The value of e is approximately

2.718281828 . . .) Log base 10, $\log_{10}$, is known as the **common logarithm** and is written as log, with the base not written but understood to be 10. Log base e, $\log_e$, is known as the **natural logarithm** and is written as ln.

Example 7: Find the following logarithms.

      (a) log 100   (c) log 0.1   (e) ln e^2

      (b) log 10,000   (d) ln e

(a)   Let   log 100 = x

                                        ↑ base 10 is understood

      Then $10^x = 100$

      Since 100 = 10^2, $10^x = 10^2$

      So $x = 2$

      Therefore,   log 100 = 2

(b)   Let   log 10,000 = x

      Then $10^x = 10{,}000$

      Since 10,000 = 10^4, $10^x = 10^4$

      So $x = 4$

      Therefore,   log 10,000 = 4

(c)   Let   log 0.1 = x

      Then $10^x = 0.1$

      Since 0.1 = 10^{-1}, $10^x = 10^{-1}$

      So $x = -1$

      Therefore,   log 0.1 = -1

(d)   Let   ln $e = x$

      Then $\log_e e = x$

      which implies $e^x = e$

      Since $e = e^1$ $e^x = e^1$

      So $x = 1$

      Therefore,   ln $e = 1$

(e)   Let $\ln e^2 = x$

Then $\log_e e^2 = x$

which implies $e^x = e^2$

So $x = 2$

Therefore, $\ln e^2 = 2$

Properties of Logarithms

The properties of logarithms assume the following about the variables M, N, b, and x.

$$M > 0, N > 0, b > 0,$$
$$b \neq 1 \qquad x \text{ is a real number}$$

1. $\log_b b = 1$
2. $\log_b 1 = 0$
3. $\log_b b^x = x$
4. $b^{\log_b x} = x$
5. $\log_b(MN) = \log_b(M) + \log_b(N)$
6. $\log_b\left(\dfrac{M}{N}\right) = \log_b M - \log_b N$

    Note: Don't confuse $\log_b\left(\dfrac{M}{N}\right)$ with $\dfrac{\log_b M}{\log_b N}$.

    To find the latter, first evaluate each log separately and then do the division.
7. $\log_b M^x = x \log_b M$
8. If $\log_b x = \log_b y$, then $x = y$.
9. $\log_b x = \dfrac{\log x}{\log b}$

    This is known as the change of base formula.

Example 8: Simplify each of the following expressions.

      (a) $\log_7 7$ (b) $\log_5 1$ (c) $\log_4 4^3$ (d) $6^{\log_6 5}$

(a) $\log_7 7 = 1$ (property 1)

(b) $\log_5 1 = 0$ (property 2)

(c) $\log_4 4^3 = 3$ (property 3)

(d) $6^{\log_6 5} = 5$ (property 4)

Example 9: If $\log_3 5 \approx 1.5$ and $\log_3 2 \approx 0.6$, approximate the following by using the properties of logarithms.

       (a) $\log_3 10$ (c) $\log_3 25$ (e) $\log_3 1.5$

       (b) $\log_3 \left(\frac{5}{2}\right)$ (d) $\log_3 \sqrt{5}$ (f) $\log_3 200$

(a) $\log_3 10 = \log_3 (5 \cdot 2)$

             $= \log_3 5 + \log_3 2$ (property 5)

             $\approx 1.5 + 0.6$

             $= 2.1$

(b) $\log_3 \left(\frac{5}{2}\right) = \log_3 5 - \log_3 2$ (property 6)

              $\approx 1.5 - 0.6$

              $= 0.9$

(c) $\log_3 25 = \log_3 5^2$

             $= 2 \log_3 5$ (property 7)

             $\approx 2(1.5)$

             $= 3$

(d) $\log_3 \sqrt{5} = \log_3 5^{1/2}$

             $= \frac{1}{2} \log_3 5$ (property 7)

             $\approx \frac{1}{2}(1.5)$

             $= .75$

(e) $\log_3 1.5 = \log_3 \left(\frac{3}{2}\right)$

             $= \log_3 3 - \log_3 2$ (property 6)

             $\approx 1 - 0.6$ (property 1, $\log_3 3 = 1$)

             $= 0.4$

(f) $\log_3 200 = \log_3 [(2^3)(5^2)]$

             $= \log_3 2^3 + \log_3 5^2$ (property 5)

             $= 3 \log_3 2 + 2 \log_3 5$ (property 7)

             $\approx 3(0.6) + 2(1.5)$

             $= 1.8 + 3$

             $= 4.8$

Example 10: Rewrite each expression as the logarithm of a single quantity.

 (a) $2 \log_b x + \frac{1}{3} \log_b y$ (b) $\frac{1}{2} \log_b(x - 2) - \log_b y + \log_b z$

(a) $\quad 2 \log_b x + \frac{1}{3} \log_b y = \log_b x^2 + \log_b y^{1/3}$ (property 7)

$\qquad\qquad\qquad\qquad = \log_b x^2 y^{1/3}$ (property 5)

$\qquad\qquad\qquad\qquad = \log_b x^2 \sqrt[3]{y}$

(b) $\quad \frac{1}{2} \log_b(x - 2) - \log_b y + 3\log_b z = \log_b(x - 2)^{1/2}$

$\qquad\qquad\qquad\qquad\qquad - \log_b y + \log_b z^3$ (property 7)

$\qquad\qquad\qquad\qquad = \log_b \dfrac{(x - 2)^{1/2}}{y} + \log_b z^3$ (property 6)

$\qquad\qquad\qquad\qquad = \log_b \dfrac{z^3 (x - 2)^{1/2}}{y}$ (property 5)

$\qquad\qquad\qquad\qquad = \log_b \dfrac{z^3 \sqrt{x - 2}}{y}$

Exponential and Logarithmic Equations

An **exponential equation** is an equation in which the variable appears in an exponent. A **logarithmic equation** is an equation that involves the logarithm of an expression containing a variable. To solve exponential equations, take the common logarithm of both sides of the equation and then apply property 7.

Example 11: Solve the following equations.

 (a) $3^x = 5$ (b) $6^{x-3} = 2$ (c) $2^{3x-1} = 3^{2x-2}$

(a) $\qquad\qquad\qquad\qquad 3^x = 5$

$\qquad\qquad\qquad\qquad \log(3^x) = \log 5$

$\qquad\qquad\qquad\qquad x \log 3 = \log 5$ (property 7)

Dividing both sides by log 3,

$$x = \frac{\log 5}{\log 3}$$

Using a calculator for approximation,

$$x \approx \frac{0.699}{0.477}$$

$$\approx 1.465$$

(b)
$$6^{x-3} = 2$$
$$\log (6^{x-3}) = \log 2$$
$$(x-3) \log 6 = \log 2 \qquad \text{(property 7)}$$

Dividing both sides by log 6,

$$x - 3 = \frac{\log 2}{\log 6}$$
$$x = \frac{\log 2}{\log 6} + 3$$

Using a calculator for approximation,

$$x \approx \frac{0.301}{0.778} + 3$$
$$\approx 3.387$$

(c)
$$2^{3x-1} = 3^{2x-2}$$
$$\log (2^{3x-1}) = \log (3^{2x-2})$$
$$(3x-1) \log 2 = (2x-2) \log 3 \quad \text{(property 7)}$$

Using the distributive property,

$$3x \log 2 - \log 2 = 2x \log 3 - 2 \log 3$$

Gathering all terms involving the variable on one side of the equation,

$$3x \log 2 - 2x \log 3 = \log 2 - 2 \log 3$$

Factoring out an x,

$$x(3 \log 2 - 2 \log 3) = \log 2 - 2 \log 3$$

Dividing both sides by 3 log 2 − 2 log 3,

$$x = \frac{\log 2 - 2 \log 3}{3 \log 2 - 2 \log 3}$$

Using a calculator for approximation,

$$x \approx 12.770$$

To solve an equation involving logarithms, use the properties of logarithms to write the equation in the form $\log_b M = N$, and then change this to exponential form, $M = b^N$.

Example 12: Solve the following equations.

$$\text{(a) } \log_4 (3x - 2) = 2$$
$$\text{(b) } \log_3 x + \log_3 (x - 6) = 3$$
$$\text{(c) } \log_2 (5 + 2x) - \log_2 (4 - x) = 3$$
$$\text{(d) } \log_5 (7x - 9) = \log_5 (x^2 - x - 29)$$

(a) $\log_4 (3x - 2) = 2$

Change to exponential form.

$$(3x - 2) = 4^2$$
$$3x - 2 = 16$$
$$3x = 18$$
$$x = 6$$

Check the answer.

$$\log_4 (3x - 2) = 2$$
$$\log_4 [3(6) - 2] \overset{?}{=} 2$$
$$\log_4 16 \overset{?}{=} 2$$
$$4^2 = 16 \checkmark$$

This is a true statement. Therefore, the solution is $x = 6$.

(b) $\log_3 x + \log_3 (x - 6) = 3$

$$\log_3 [x(x - 6)] = 3 \qquad \text{(property 5)}$$

Change to exponential form.

$$x(x - 6) = 3^3$$
$$x^2 - 6x = 27$$
$$x^2 - 6x - 27 = 0$$
$$(x - 9)(x + 3) = 0$$
$$x - 9 = 0 \qquad \text{or} \qquad x + 3 = 0$$
$$x = 9 \qquad\qquad\qquad x = -3$$

Check the answers.

If $x = 9$	If $x = -3$
$\log_3 x + \log_3(x-6) = 3$	$\log_3 x + \log_3(x-6) = 3$
$\log_3 9 + \log_3(9-6) \overset{?}{=} 3$	$\log_3(-3) + \log_3(-3-6) \overset{?}{=} 3$ No
$2 + 1 = 3$ ✓	

Since the logarithm of a negative number is not defined, the only solution is $x = 9$.

(c) $\log_2(5 + 2x) - \log_2(4 - x) = 3$

$$\log_2\left(\frac{5+2x}{4-x}\right) = 3 \qquad \text{(property 6)}$$

Change into exponential form.

$$\frac{5+2x}{4-x} = 2^3$$

$$\frac{5+2x}{4-x} = 8$$

Using the cross products property,

$$5 + 2x = 8(4 - x)$$
$$5 + 2x = 32 - 8x$$
$$10x = 27$$
$$x = 2.7$$

Check the answer.

$$\log_2(5 + 2x) - \log_2(4 - x) = 3$$

$$\log_2[5 + 2(2.7)] - \log_2[4 - 2.7] \overset{?}{=} 3$$

$$\log_2(10.4) - \log_2(1.3) \overset{?}{=} 3$$

$$\log_2(10.4/1.3) \overset{?}{=} 3$$

$$\log_2(8) \overset{?}{=} 3$$

$$2^3 = 8 \checkmark$$

This is a true statement. Therefore, the solution is $x = 2.7$.

(d) $\log_5 (7x - 9) = \log_5 (x^2 - x - 29)$

$$7x - 9 = x^2 - x - 29 \qquad \text{(property 8)}$$
$$0 = x^2 - 8x - 20$$
$$0 = (x - 10)(x + 2)$$

$x - 10 = 0 \qquad \text{or} \qquad x + 2 = 0$

$\qquad x = 10 \qquad\qquad\qquad x = -2$

Check the answers.

If $x = 10$,

$$\log_5(7x - 9) = \log_5(x^2 - x - 29)$$
$$\log_5[7(10) - 9] \overset{?}{=} \log_5[10^2 - 10 - 29]$$
$$\log_5(61) = \log_5(61) \checkmark$$

This is a true statement.

If $x = -2$,

$$\log_5(7x - 9) = \log_5(x^2 - x - 29)$$
$$\log_5[7(-2) - 9] \overset{?}{=} \log_5[(-2)^2 - (-2) - 29]$$
$$\log_5(-23) \overset{?}{=} \log_5(-23)$$

This appears to be true, but $\log_5(-23)$ is not defined. Therefore, the only solution is $x = 10$.

Example 13: Find $\log_3 8$.

$$\log_3 8 = \frac{\log 8}{\log 3} \qquad \text{(property 9, the change of base formula)}$$

Note: $\log 8 = \log_{10} 8$ and $\log 3 = \log_{10} 3$.

Using a calculator for approximation,

$$\log_3 8 \approx \frac{0.903}{0.477}$$
$$\approx 1.893$$

Chapter Checkout

1. Rewrite each equation in logarithmic form.

$$\text{a) } 6^2 = 36 \quad \text{b) } 3^{-3} = \tfrac{1}{27} \quad \text{c) } \left(\tfrac{1}{5}\right)^{-3} = 125$$

2. Solve the following equations if possible:

$$\text{a) } \log_9 81 = x \quad \text{b) } \log_2\left(\tfrac{1}{16}\right) = x \quad \text{c) } \log_9 y = -3$$

3. Find the following common logarithms:

$$\text{a) } \log 1000 \quad \text{b) } \log .01 \quad \text{c) } \log 100,000$$

Answers: 1. a. $2 = \log_6 36$ b. $-3 = \log_3\left(\tfrac{1}{27}\right)$ c. $-3 = \log_{\left(\tfrac{1}{5}\right)} 125$
2. a. $x = 2$ b. $x = -4$ c. $y = \tfrac{1}{729}$ 3. a. 3 b. -2 c. 5

Chapter 15

SEQUENCES AND SERIES

Chapter Check-In

❏ Studying arithmetic sequences

❏ Studying arithmetic series

❏ Studying geometric sequences

❏ Studying geometric series

❏ Using summation notation

Arithmetic (with the emphasis on the 1st and 3rd syllables) sequences are sequences that grow with regular spacing. 2, 4, 6, 8, . . . is an example. Notice that each number is two higher than the one before it (in this example, that is). A geometric sequence involves multiplication by a single number, for example, 3, 12, 48, 192, ... Notice the multiplier here is 4. Also notice that a geometric sequence grows (or diminishes) much more rapidly than an arithmetic one.

Series are sums of sequences with a definite number of terms. Within this chapter, arithmetic and geometric series, as well as sequences and summation notation, are studied.

Definition and Examples of Sequences

A **sequence** is an ordered list of numbers.

$$\left.\begin{array}{l} 1, 3, 5, 7, 9, \ldots \\ -8, 3, 14, 25, \ldots \\ 1, 2, 3, 8, 16, \ldots \end{array}\right\} \quad \text{are examples of sequences}$$

The three dots mean to continue on in the pattern established. Each number in the sequence is called a **term.** In the sequence 1, 3, 5, 7, 9, . . . , 1 is the first term, 3 is the second term, 5 is the third term, and so on. The notation a_1, a_2, a_3, . . . , a_n is used to denote the different terms in a sequence. The expression a_n is referred to as the **general** or **nth term** of the sequence.

Example 1: Write the first five terms of a sequence described by the general term $a_n = 3n + 2$.

$$a_n = 3n + 2$$
$$a_1 = 3(1) + 2 = 5$$
$$a_2 = 3(2) + 2 = 8$$
$$a_3 = 3(3) + 2 = 11$$
$$a_4 = 3(4) + 2 = 14$$
$$a_5 = 3(5) + 2 = 17$$

Therefore, the first five terms are 5, 8, 11, 14, and 17.

Example 2: Write the first five terms of $a_n = 2(3^{n-1})$.

$$a_n = 2(3^{n-1})$$
$$a_1 = 2(3^{1-1}) = 2(3^0) = 2(1) = 2$$
$$a_2 = 2(3^{2-1}) = 2(3^1) = 2(3) = 6$$
$$a_3 = 2(3^{3-1}) = 2(3^2) = 2(9) = 18$$
$$a_4 = 2(3^{4-1}) = 2(3^3) = 2(27) = 54$$
$$a_5 = 2(3^{5-1}) = 2(3^4) = 2(81) = 162$$

Therefore, the first five terms are 2, 6, 18, 54, and 162.

Example 3: Find an expression for the *n*th term of each sequence.

    (a) 2, 4, 6, 8, . . .
    (b) 10, 50, 250, 1250, . . .
    (c) 3, 7, 11, 15, 19, . . .

(a) 2, 4, 6, 8, . . . $a_1 = 2 = 2(1)$
$$a_2 = 4 = 2(2)$$
$$a_3 = 6 = 2(3)$$
$$a_4 = 8 = 2(4)$$

Based on this pattern, $a_n = 2n$.

(b) 10, 50, 250, 1250, . . .

$a_1 = 10 = 2(5) = 2(5^1)$

$a_2 = 50 = 2(25) = 2(5^2)$

$a_3 = 250 = 2(125) = 2(5^3)$

$a_4 = 1250 = 2(625) = 2(5^4)$

Based on this pattern, $a_n = 2(5^n)$.

(c) 3, 7, 11, 15, 19, . . .

$a_1 = 3$

$a_2 = 7 = 3 + 4(1)$

$a_3 = 11 = 3 + 8 = 3 + 4(2)$

$a_4 = 15 = 3 + 12 = 3 + 4(3)$

$a_5 = 19 = 3 + 16 = 3 + 4(4)$

Based on this pattern,

$$a_n = 3 + 4(n-1)$$
$$= 3 + 4n - 4$$
$$= 4n - 1$$

Arithmetic Sequence

An **arithmetic sequence** is a sequence in which, beginning with the second term, each term is found by adding the same value to the previous term. Its general term is described by

$$a_n = a_1 + (n-1)d$$

The number d is called the **common difference.** It can be found by taking any term in the sequence and subtracting its preceding term.

Example 4: Find the common difference in each of the following arithmetic sequences. Then express each sequence in the form $a_n = a_1 + (n-1)d$ and find the twentieth term of the sequence.

(a) 1, 5, 9, 13, 17, . . .

(b) $-\frac{5}{8}, -\frac{3}{8}, -\frac{1}{8}, \frac{1}{8}, \frac{3}{8}$ . . .

(a) 1, 5, 9, 13, 17, . . .

Since $\qquad d = a_2 - a_1 = 5 - 1 = 4$

Then $\qquad a_n = 1 + (n-1)4$

Therefore, the twentieth term of the sequence is

$$a_{20} = 1 + (20 - 1)4 = 1 + 76 = 77$$

(b) $-\frac{5}{8}, -\frac{3}{8}, -\frac{1}{8}, \frac{1}{8}, \frac{3}{8} \ldots$

Since $\qquad d = a_2 - a_1 = -\frac{3}{8} - \left(-\frac{5}{8}\right) = \frac{1}{4}$

Then $\qquad a_n = -\frac{5}{8} + (n - 1)\left(\frac{1}{4}\right)$

Therefore, the twentieth term of the sequence is

$$a_{20} = -\frac{5}{8} + (20 - 1)\left(\frac{1}{4}\right) = -\frac{5}{8} + \frac{19}{4} = \frac{33}{8}$$

Arithmetic Series

An **arithmetic series** is the sum of the terms in an arithmetic sequence with a definite number of terms. There is a simple formula for finding the sum. If S_n represents the sum of an arithmetic sequence with terms a_1, a_2, a_3, . . . , a_n, then

$$(1) \ \ S_n = \frac{n}{2}(a_1 + a_n)$$

This formula requires the values of the first and last terms and the number of terms.

Since $\qquad\qquad a_n = a_1 + (n - 1)d$

Then $\qquad\qquad a_1 + a_n = a_1 + [a_1 + (n - 1)d]$

$$a_1 + a_n = 2a_1 + (n - 1)d$$

Substituting this last expression for $(a_1 + a_n)$ into formula (1) above, another formula for the sum of an arithmetic sequence is formed.

$$(2) \ \ S_n = \frac{n}{2}[2a_1 + (n - 1)d]$$

This formula for the sum of an arithmetic sequence requires the first term, the common difference, and the number of terms.

Example 5: In the arithmetic sequence -3, 4, 11, 18, . . . , find the sum of the first twenty terms.

$$-3, 4, 11, 18, \ldots$$

$$a_1 = -3$$

$$d = 4 - (-3) = 7$$

$$n = 20$$

Use formula (2) to find the sum.

$$S_{20} = \frac{n}{2}[2a_1 + (n-1)d]$$
$$= \frac{20}{2}[2(-3) + (20-1)(7)]$$
$$= 10(-6 + 133)$$
$$= 1270$$

Example 6: Find the sum of the multiples of 3 between 28 and 112.

The first multiple of 3 between 28 and 112 is 30, and the last multiple of 3 between 28 and 112 is 111. In order to use formula (1), the number of terms must be known. $a_n = a_1 + (n-1)d$ can be used to find n.

$$a_n = 111, \quad a_1 = 30, \quad d = 3$$
$$111 = 30 + (n-1)(3)$$
$$81 = (n-1)(3)$$
$$27 = (n-1)$$
$$28 = n$$

Now, use formula (1).

$$S_n = \frac{n}{2}(a_1 + a_2)$$
$$S_{28} = \frac{28}{2}(30 + 111)$$
$$= 14(141)$$
$$= 1974$$

The sum of the multiples of 3 between 28 and 112 is 1974.

Geometric Sequence

A **geometric sequence** is a sequence in which each term is found by multiplying the same value to the previous term. Its general term is

$$a_n = a_1 r^{n-1}$$

The value r is called the **common ratio.** It is found by taking any term in the sequence and dividing it by its preceding term.

Example 7: Find the common ratio in each of the following geometric sequences. Then express each sequence in the form $a_n = a_1 r^{n-1}$ and find the eighth term of the sequence.

$$\text{(a) } 1, 3, 9, 27, \ldots$$
$$\text{(b) } 64, -16, 4, -1, \ldots$$
$$\text{(c) } 16, 24, 36, 54, \ldots$$

(a) 1, 3, 9, 27, . . .

Since $\qquad\qquad r = \frac{a_2}{a_1} = \frac{3}{1} = 3$

Then $\qquad\qquad a_n = 1(3^{n-1})$

Therefore, the eighth term of the sequence is
$$a_8 = 1(3^{8-1})$$
$$= 1(3^7)$$
$$= 2187$$

(b) 64, −16, 4, −1, . . .

Since $\qquad\qquad r = \frac{a_2}{a_1} = \frac{-16}{64} = -\frac{1}{4}$

Then $\qquad\qquad a_n = 64\left(-\frac{1}{4}\right)^{n-1}$

Therefore, the eighth term of the sequence is
$$a_8 = 64\left(-\frac{1}{4}\right)^{8-1}$$
$$= 64\left(-\frac{1}{4}\right)^7$$
$$= -\frac{1}{256}$$

(c) 16, 24, 36, 54, . . .

Since $\qquad\qquad r = \frac{a_2}{a_1} = \frac{24}{16} = \frac{3}{2}$

Then $\qquad\qquad a_n = 16\left(\frac{3}{2}\right)^{n-1}$

Therefore, the eighth term of the sequence is
$$a_8 = 16\left(\frac{3}{2}\right)^{8-1}$$
$$= 16\left(\frac{3}{2}\right)^7$$
$$= \frac{2187}{8}$$

Geometric Series

A **geometric series** is the sum of the terms in a geometric sequence. If the sequence has a definite number of terms, the simple formula for the sum is

(3) $S_n = \dfrac{a_1\left(1 - r^n\right)}{1 - r}$ $\quad (r \neq 1)$

This form of the formula is used when the number of terms (n), the first term (a_1), and the common ratio (r) are known.

Another formula for the sum of a geometric sequence is

(4) $S_n = \dfrac{a_1 - a_n r}{1 - r}$ $\quad (r \neq 1)$

This form requires the first term (a_1), the last term (a_n), and the common ratio (r) but does *not* require the number of terms (n).

Example 8: Find the sum of the first five terms of the geometric sequence in which $a_1 = 3$ and $r = -2$.

$$a_1 = 3, \ r = -2$$

Use formula (3).

$$S_n = \frac{a_1\left(1 - r^n\right)}{1 - r}$$

$$S_5 = \frac{3\left[1 - (-2)^5\right]}{1 - (-2)}$$

$$= \frac{3\left[1 - (-32)\right]}{3}$$

$$= \frac{99}{3}$$

$$= 33$$

Example 9: Find the sum of the geometric sequence for which $a_1 = 48$, $a_n = 3$, $r = -\frac{1}{2}$.

$$a_1 = 48, \ a_n = 3, \ r = -\frac{1}{2}$$

Use formula (4).

$$S_n = \frac{a_1 - a_n r}{1 - r}$$

$$= \frac{48 - 3\left(-\frac{1}{2}\right)}{1 - \left(-\frac{1}{2}\right)}$$

$$= \frac{48 + \frac{3}{2}}{\frac{3}{2}}$$

$$= \frac{\frac{99}{2}}{\frac{3}{2}}$$

$$= 33$$

Example 10: Find a_1 in each geometric series described.

(a) $S_n = 244$, $r = -3$, $n = 5$

(b) $S_n = 15.75$, $r = 0.5$, $a_n = 0.25$

(a) $S_n = 244$, $r = -3$, $n = 5$

Use formula (3).

$$S_n = \frac{a_1\left(1 - r^n\right)}{1 - r}$$

$$244 = \frac{a_1\left[1 - (-3)^5\right]}{1 - (-3)}$$

$$244 = \frac{a_1(244)}{4}$$

$$a_1 = 4$$

(b) $S_n = 15.75$, $r = 0.5$, $a_n = 0.25$

Use formula (4).

$$S_n = \frac{a_1 - a_n r}{1 - r}$$

$$15.75 = \frac{a_1 - (0.25)(0.5)}{1 - 0.5}$$

$$15.75 = \frac{a_1 - 0.125}{0.5}$$

$$7.875 = a_1 - 0.125$$

$$a_1 = 8$$

If a geometric series is infinite (that is, endless) and $-1 < r < 1$, then the formula for its sum becomes

$$(5) \quad S = \frac{a_1}{1 - r}$$

If $r > 1$ or if $r < -1$, then the infinite series does not have a sum.

Example 11: Find the sum of each of the following geometric series.

$$(a) \; 25 + 20 + 16 + 12.8 + \ldots$$

$$(b) \; 3 - 9 + 27 - 81 + \ldots$$

(a) $25 + 20 + 16 + 12.8 + \ldots$

First find r. $\qquad\qquad\qquad r = \frac{20}{25} = \frac{4}{5}$

Since $-1 < \frac{4}{5} < 1$, this infinite geometric series has a sum. Use formula (5).

$$S = \frac{a_1}{1 - r}$$

$$= \frac{25}{1 - \frac{4}{5}}$$

$$= \frac{25}{\frac{1}{5}}$$

$$= 125$$

(b) $3 - 9 + 27 - 81 + \ldots$

First find r. $\qquad\qquad\qquad r = \frac{-9}{3} = -3$

Since $-3 < -1$, this geometric series does not have a sum.

Summation Notation

A simple method for indicating the sum of a finite (ending) number of terms in a sequence is the **summation notation.** This involves the Greek letter sigma, Σ. When using the sigma notation, the variable defined below the Σ is called the **index of summation.** The lower number is the lower limit of the index (the term where the summation starts), and the upper number is the upper limit of the summation (the term where the summation ends). Consider

$$\sum_{k=2}^{7}(2k + 3)$$

This is read as "the summation of $(2k + 3)$ as k goes from 2 to 7." The replacements for the index are always consecutive integers.

$$\sum_{k=2}^{7}(2k+3) = \overset{k=2}{\left[2(2)+3\right]} + \overset{k=3}{\left[2(3)+3\right]} + \overset{k=4}{\left[2(4)+3\right]} + \overset{k=5}{\left[2(5)+3\right]}$$

$$+ \overset{k=6}{\left[2(6)+3\right]} + \overset{k=7}{\left[2(7)+3\right]}$$

$$= 7 + 9 + 11 + 13 + 15 + 17$$

$$= 72$$

Example 12: Write out the terms of the following sums; then compute the sum.

$$\text{(a) } \sum_{j=0}^{5}3j \quad \text{(b) } \sum_{k=0}^{4}2^{k} \quad \text{(c) } \sum_{n=1}^{5}\frac{3n+2}{n+1}$$

(a) $\sum_{j=0}^{5}3j = \overset{j=0}{3(0)} + \overset{j=1}{3(1)} + \overset{j=2}{3(2)} + \overset{j=3}{3(3)} + \overset{j=4}{3(4)} + \overset{j=5}{3(5)} = 45$

(b) $\sum_{k=0}^{4}2^{k} = \overset{k=0}{2^0} + \overset{k=1}{2^1} + \overset{k=2}{2^2} + \overset{k=3}{2^3} + \overset{k=4}{2^4} = 31$

(c) $\sum_{n=1}^{5}\frac{3n+2}{n+1} = \overset{n=1}{\frac{5}{2}} + \overset{n=2}{\frac{8}{3}} + \overset{n=3}{\frac{11}{4}} + \overset{n=4}{\frac{14}{5}} + \overset{n=5}{\frac{17}{6}}$

$$= \frac{150 + 160 + 165 + 168 + 170}{60}$$

$$= \frac{813}{60}$$

$$= \frac{271}{20}$$

Example 13: Use sigma notation to express each series.

(a) $8 + 11 + 14 + 17 + 20$ (b) $\frac{2}{3} - 1 + \frac{3}{2} - \frac{9}{4} + \frac{27}{8} - \frac{81}{16}$

(a) $8 + 11 + 14 + 17 + 20$

This is an arithmetic series with five terms whose first term is 8 and whose common difference is 3. Therefore, $a_1 = 8$ and $d = 3$. The nth term of the corresponding sequence is

$$a_n = a_1 + (n-1)d$$

$$= 8 + (n-1)3$$

$$= 3n + 5$$

Since there are five terms, the given series can be written as

$$\sum_{n=1}^{5} a_n = \sum_{n=1}^{5}(3n+5)$$

(b) $\frac{2}{3} - 1 + \frac{3}{2} - \frac{9}{4} + \frac{27}{8} - \frac{81}{16}$

This is a geometric series with six terms whose first term is $\frac{2}{3}$ and whose common ratio is $\frac{-3}{2}$. Therefore, $a_1 = \frac{2}{3}$ and $r = \frac{-3}{2}$. The nth term of the corresponding sequence is

$$a_n = a_1 r^{n-1}$$
$$= \frac{2}{3}\left(\frac{-3}{2}\right)^{n-1}$$

Since there are six terms in the given series, the sum can be written as

$$\sum_{n=1}^{6} a_n = \sum_{n=1}^{6} \frac{2}{3}\left(\frac{-3}{2}\right)^{n-1}$$

Chapter Checkout

1. Find the twentieth term of each arithmetic sequence:

　　　a) $0, 6, 12, \ldots$　　b) $-\frac{3}{4}, -\frac{2}{4}, -\frac{1}{4}, \ldots$　　c) $5, 4.4, 3.8, \ldots$

2. Find the sum of multiples of 5 between 4 and 112.

3. Find the 8th term of the following geometric sequences:

　　　a) $5, -15, 45, \ldots$
　　　b) $4, 20, 100, \ldots$

4. Write out the terms of the sums. Then compute the sums.

$$\text{a) } \sum_{n=2}^{9}(3n-5) \quad \text{b) } \sum_{w=0}^{6} 3^w$$

Answers:
1. a. 114　b. 4　c. −6.4　　2. 1265　　3. a. −10,935　b. 312,500
4. a. 1, 4, 7, 10, 13, 16, 19, 22; 92　　b. 1, 3, 9, 27, 81, 243, 729; 1093

Chapter 16

ADDITIONAL TOPICS

Chapter Check-In

❑ Expressing factorials as multiplication

❑ Using the binomial theorem and coefficients

❑ Applying permutations

❑ Distinguishing combinations from permutations

This chapter examines several topics that do not neatly fit into the overall scheme of topics of Algebra II. The first to be examined is factorials — a notation that expresses a natural number multiplied together by all its predecessors. 4! (read 4 factorial), for example, means $4 \times 3 \times 2 \times 1$. Factorials play a part in the computing of permutations and combinations. Both play a part in probability, and both are examined within this chapter.

Additionally, the topics of binomial theorem and the patterns formed by exponential expansion of binomials — the most notable being Pascal's triangle — are explored herein.

Factorials

Factorial is a convenient way of expressing the answer to any natural number multiplied together with all its preceding natural numbers. The symbol for factorial is !. 6! is read as "six factorial."

$$6! = (6)(5)(4)(3)(2)(1) = 720$$

Since zero is not a natural number, 0! would have no meaning. So that the expression 0! can be used to express formulas, it is defined as having the value of 1.

$$n! = n(n-1)(n-2)(n-3) \ldots (1)$$

$$0! = 1$$

Example 1: Evaluate $\dfrac{9!}{3!6!}$.

$$\frac{9!}{3!6!} = \frac{(9)(8)(7)(\cancel{6})(\cancel{5})(\cancel{4})(\cancel{3})(\cancel{2})(\cancel{1})}{(3)(2)(1)(\cancel{6})(\cancel{5})(\cancel{4})(\cancel{3})(\cancel{2})(\cancel{1})} = 84$$

Binomial Coefficients and the Binomial Theorem

When a binomial is raised to whole number powers, the coefficients of the terms form a pattern.

$$(a + b)^0 = 1$$
$$(a + b)^1 = 1a + 1b$$
$$(a + b)^2 = 1a^2 + 2ab + 1b^2$$
$$(a + b)^3 = 1a^3 + 3a^2b + 3ab^2 + 1b^3$$
$$(a + b)^4 = 1a^4 + 4a^3b + 6a^2b^2 + 4ab^3 + 1b^4$$
$$(a + b)^5 = 1a^5 + 5a^4b + 10a^3b^2 + 10a^2b^3 + 5ab^4 + 1b^5$$

These expressions exhibit many patterns:

■ Each expansion has one more term than the power on the binomial.

■ The sum of the exponents in each term in the expansion is the same as the power on the binomial.

■ The powers on a in the expansion decrease by 1, while the powers on b increase by 1.

■ The coefficients form a symmetrical pattern.

■ Each entry below the second row is the sum of the closest pair of numbers in the line directly above it.

This triangular array is called **Pascal's triangle,** named after the French mathematician Blaise Pascal.

Pascal's triangle can be extended to find the coefficients for raising a binomial to any whole number exponent. This same array could be expressed using the factorial symbol, as shown in the following.

row ⓪	1 ⟶	$\frac{0!}{0!}$
row ①	1 1 ⟶	$\frac{1!}{0!1!}$ $\frac{1!}{1!0!}$
row ②	1 2 1 ⟶	$\frac{2!}{0!2!}$ $\frac{2!}{1!1!}$ $\frac{2!}{2!0!}$
row ③	1 3 3 1 ⟶	$\frac{3!}{0!3!}$ $\frac{3!}{1!2!}$ $\frac{3!}{2!1!}$ $\frac{3!}{3!0!}$
row ④	1 4 6 4 1 ⟶	$\frac{4!}{0!4!}$ $\frac{4!}{1!3!}$ $\frac{4!}{2!2!}$ $\frac{4!}{3!1!}$ $\frac{4!}{4!0!}$
row ⑤	1 5 10 10 5 1 ⟶	$\frac{5!}{0!5!}$ $\frac{5!}{1!4!}$ $\frac{5!}{2!3!}$ $\frac{5!}{3!2!}$ $\frac{5!}{4!1!}$ $\frac{5!}{5!0!}$
row ⓝ	1..................1 →	$\frac{n!}{0!n!}$ $\frac{n!}{1!(n-1)!}$ .... $\frac{n!}{(n-1)!1!}$ $\frac{n!}{n!0!}$

In general,

$$(a+b)^n = \frac{n!}{0!n!}\left(a^n\right)b^0 + \frac{n!}{1!(n-1)!}\left(a^{n-1}\right)b^1 + \frac{n!}{2!(n-2)!}\left(a^{n-2}\right)b^2$$

$$+ \frac{n!}{3!(n-3)!}\left(a^{n-3}\right)b^3 + ... + \frac{n!}{n!0!}a^0 b^n$$

The symbol $\binom{n}{k}$, called the **binomial coefficient,** is defined as follows:

$$\binom{n}{k} = \frac{n!}{k!(n-k)!}$$

Therefore,

$$(a+b)^n = \binom{n}{0}a^n b^0 + \binom{n}{1}a^{n-1}b^1 + \binom{n}{2}a^{n-2}b^2 + ... + \binom{n}{n}a^0 b^n$$

This could be further condensed using sigma notation.

$$(a+b)^n = \sum_{k=0}^{n}\binom{n}{k}a^{n-k}b^k$$

This formula is known as the **binomial theorem.**

Example 2: Use the binomial theorem to express $(x+y)^7$ in expanded form.

$$(x+y)^7 = \sum_{k=0}^{7}\binom{7}{k}x^{7-k}y^k$$

$$= \binom{7}{0}x^7 y^0 + \binom{7}{1}x^6 y^1 + \binom{7}{2}x^5 y^2 + \binom{7}{3}x^4 y^3 + \binom{7}{4}x^3 y^4$$

$$+ \binom{7}{5}x^2 y^5 + \binom{7}{6}x^1 y^6 + \binom{7}{7}x^0 y^7$$

$$= \frac{7!}{0!7!} x^7 y^0 + \frac{7!}{1!6!} x^6 y + \frac{7!}{2!5!} x^5 y^2 + \frac{7!}{3!4!} x^4 y^3 + \frac{7!}{4!3!} x^3 y^4$$

$$+ \frac{7!}{5!2!} x^2 y^5 + \frac{7!}{6!1!} xy^6 + \frac{7!}{7!0!} x^0 y^7$$

$$= x^7 + 7x^6 y + 21x^5 y^2 + 35x^4 y^3 + 35x^3 y^4 + 21x^2 y^5 + 7xy^6 + y^7$$

Notice the following pattern:

■ the first term $= \binom{7}{0} x^7 y^0$

■ the second term $= \binom{7}{1} x^6 y^1$

■ the third term $= \binom{7}{2} x^5 y^2$

In general, the kth term of any binomial expansion can be expressed as follows:

$$k\text{th term} = \binom{n}{k-1} x^{n-(k-1)} y^{k-1}$$

Example 3: Find the tenth term of the expansion $(x + y)^{13}$

$$k\text{th term} = \left(\frac{n}{k-1} \right) x^{n-(k-1)} y^{k-1}$$

Since $n = 13$ and $k = 0$,

$$10\text{th term} = \binom{13}{9} x^{13-9} y^9$$

$$= \frac{13!}{9!4!} x^4 y^9$$

$$= 715 x^4 y^9$$

Permutations

Multiplication principle for events: If one event can occur in p different ways, and another independent event can occur in q different ways, then there are pq ways that both events can occur together.

Example 4: How many different ways can a man coordinate a wardrobe if he has a choice of four different pants and two different sport jackets?

The first event (selecting pants) can occur in four different ways. The second event (selecting a sport jacket) can occur in two ways. Therefore, according to the multiplication principle for events, there are $(4)(2) = 8$ different ways to select a wardrobe.

Example 5: In how many ways can five books be arranged on a shelf?

The first space can be filled with any of the five books, the second space with any of the remaining four books, the third space with any of the remaining three books, the fourth space with either of the remaining two books, and the fifth space with the last book. Therefore, there are $(5)(4)(3)(2)(1) = 120$ ways to arrange five books on a shelf.

Example 6: In how many ways can three out of eight books be arranged on a shelf?

The first space can be filled in any of eight ways, the second space in any of the remaining seven ways, the third space in any of the remaining six ways. Therefore, there are $(8)(7)(6) = 336$ ways that three out of eight books can be arranged on a shelf.

The arrangement of objects in a certain order is called a **permutation.** The number of ways to arrange eight things taken three at a time is written as $P(8, 3)$ or as $_8P_3$. $P(n, r)$ or $_nP_r$ is read as "the permutation of n things taken r at a time."

$$_nP_r = P(n, r) = \frac{n!}{(n-r)!}$$

If any of the objects in a permutation are repeats, a different formula is used. The number of permutations of n objects of which p are alike and q are alike is

$$\frac{n!}{p!q!}$$

Example 7: How many different ways can the letters in the word "Mississippi" be arranged?

"Mississippi" has eleven letters. Since "i" is repeated four times, "s" is repeated four times, and "p" is repeated two times, then there are

$$\frac{11!}{4!4!2!} = 34,650$$

different ways that the letters in the word "Mississippi" can be arranged.

Combinations

When the order in which objects are chosen is not important, as it is in a permutation, the arrangement is called a **combination.** The combination of n things taken r at a time is written as $C(n, r)$ or $_nC_r$. A formula for finding the number of combinations of n objects taken r at a time is given by the following:

$$_nC_r = C(n, r) = \frac{n!}{r!(n-r)!}$$

Note that this is the same as the binomial coefficient formula.

Example 8: A class has sixteen students. How many groups consisting of four students can be formed?

This is a combinations problem, since the order in which the students in a group are chosen is not important. In this problem, $n = 16$ and $r = 4$.

$$
\begin{aligned}
C(16, 4) &= \frac{16!}{4!(16-4)!} \\
&= \frac{(16)(15)(14)(13)\,\cancel{(12!)}}{(4)(3)(2)(1)\,\cancel{(12!)}} \\
&= 1820
\end{aligned}
$$

There are 1820 different ways to form the groups.

Example 9: A committee in Congress consists of nine Republicans and six Democrats. In how many ways can a subcommittee be chosen if it must contain four Republicans and three Democrats?

There are $C(9, 4)$ ways of choosing the four Republicans out of the nine Republicans. There are $C(6, 3)$ ways of choosing the three Democrats out of the six Democrats. By the multiplication principle of events, there are $C(9, 4) \cdot C(6, 3)$ ways of choosing the subcommittee.

$$
\begin{aligned}
C(9, 4) \cdot C(6, 3) &= \frac{9!}{4!(9-4)!} \cdot \frac{6!}{3!(6-3)!} \\
&= \frac{(9)(8)(7)(6)(5!)}{(4)(3)(2)(1)(5!)} \cdot \frac{(6)(5)(4)(3!)}{(3)(2)(1)(3!)} \\
&= (126)(20) \\
&= 2520
\end{aligned}
$$

There are 2520 different possible subcommittees.

Chapter Checkout

1. Evaluate the following:

$$\frac{7!}{3!5!}$$

2. In how many ways can 7 bicycles be arranged in a bike rack with 7 slots?

3. In how many ways can 4 of 7 cars be parked in a driveway that has 4 parking spaces?

4. A bridge club has 20 members. How many different foursomes may be formed?

Answers: 1. 7   2. 5040   3. 840   4. 4845

Chapter 17
WORD PROBLEMS

Chapter Check-In

❑ Learning strategies for word problem solutions

❑ Solving simple interest problems

❑ Handling mixture problems

❑ Solving motion problems

❑ Solving work and series problems

If there is any answer to the question, "What good will Algebra II do me in the real world?" this chapter has it. In many occupations, from carpentry to pharmacology, from nursing to aviation, and many other fields, word problems may be a part of everyday life or may be encountered on occasion. Whichever the case, your understanding how to deal with such problems may prove to be invaluable.

This chapter explains the basic strategies to use when approaching word problems, as well as specific tactics and formulas that apply to specific types of problems.

General Strategy

Although word problems differ from one another, there is a general strategy that can be used to solve them.

1. **Read the problem carefully, several times if necessary, to familiarize yourself with what is being asked.**

2. **Select a variable or variables to represent the unknown(s) in the problem.**

3. **If possible, draw a diagram to illustrate the facts in a problem or create a chart to organize the given information.**

4. **Translate the problem into an algebraic sentence using the variable(s) chosen.**
5. **Solve the algebraic sentence.**
6. **Check your result.**
7. **Reread the problem; see how the answer to the algebraic sentence can be used to answer the question asked. Check to see if the answer is reasonable.**
8. **Finally, answer the question.**

Simple Interest

The formula necessary to solve simple interest problems is

$$I = PRT$$

where I = interest earned

P = principal (the amount invested)

R = annual interest rate (usually given as a percent)

T = length of time of the investment in years

Example 1: Jim has $10,000 to invest. He will invest part at 9% annual interest and the remaining part at 12% annual interest. After one year, he expects to earn $165 more from the 9% investment than from the 12% investment. How much will he invest at each rate?

Let x = amount invested at 9%. Then

$$0.09x = \text{amount of annual interest at 9\%}$$
$$10,000 - x = \text{amount invested at 12\%}$$

Then $0.12(10,000 - x)$ = amount of annual interest at 12%

The sentence "he expects to earn $165 more from the 9% investment than from the 12% investment" is translated into the following.

Interest at 9% is $165 more than interest at 12%

$$0.09x = 0.12(10,000 - x) + 165$$
$$0.09x = 1200 - 0.12x + 165$$
$$0.21x = 1365$$
$$x = 6500$$

The check is left to you. Jim will invest $6500 at 9% and $3500 at 12%.

Compound Interest

The formula necessary to solve most compound interest problems is

$$A = P\left(1 + \frac{r}{n}\right)^{nt}$$

where A = future value of the investment

P = principal (original investment)

r = annual interest rate

t = time in years of the investment

n = number of times the investment is compounded annually

Example 2: How long would it take for an investment of \$3500 to become \$4200 if it is invested in an account that earns 6% compounded monthly?

$A = 4200$ $P = 3500$ $r = 0.06$ $n = 12$

Therefore $A = P\left(1 + \frac{r}{n}\right)^{nt}$

becomes

$$4200 = 3500(1 + \frac{0.06}{12})^{12t}$$
$$4200 = 3500(1 + 0.005)^{12t}$$
$$4200 = 3500(1.005)^{12t}$$
$$\frac{4200}{3500} = (1.005)^{12t}$$
$$1.2 = (1.005)^{12t}$$

Since, in this problem, the variable is in the exponent, logarithms will be used to solve it.

$$\log(1.2) = \log(1.005)^{12t}$$
$$\log(1.2) = 12t[\log(1.005)]$$
$$\frac{\log(1.2)}{12\log(1.005)} = t$$
$$3.05 \approx t$$

The \$3500 investment would have become \$4200 in about 3.05 years, or just over 3 years and 2 weeks.

Mixture

Example 3: A radiator's capacity is sixteen gallons. If the radiator is full of a 40% antifreeze solution, how many gallons must be drained and replaced with a 75% solution to obtain a full radiator of 45% antifreeze solution?

A diagram is helpful here (see Figure 17-1).

Figure 17-1   A diagram of the problem posed in Example 3.

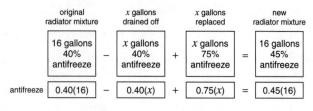

$$0.40(16) - 0.40x + 0.75x = 0.45(16)$$
$$6.4 + 0.35x = 7.2$$
$$0.35x = 0.8$$
$$x = \frac{80}{35} = \frac{16}{7}$$
$$\approx 2.29 \text{ gallons}$$

So $\frac{16}{7}$, or $2\frac{2}{7}$, gallons must be drained and replaced.

Motion

The basic formula for motion problems is

$$(r) \quad \times \quad (t) \quad = \quad d \quad [\text{or} \quad d \quad = \quad (r) \quad \times \quad (t)]$$

(rate) times (time) = distance   [or distance equals (rate) times (time)]

Example 4: A train leaves Chicago at 11:00 a.m. traveling east at a speed of 40 mph. Two hours later, a second train leaves Chicago on a parallel track traveling in the same direction as the first train. The second train travels at a rate of 50 mph. Assuming that neither train stops, at what time will the second train catch up to the first train? At that time, how far has each train traveled?

Drawing a diagram (see Figure 17-2) helps in understanding the situation. Constructing a chart helps in organizing the data.

Figure 17-2 A diagram for the problem posed in Example 4.

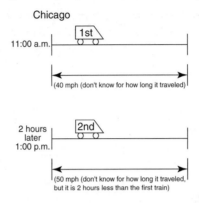

Chicago

Since both trains began at the same point and will be at the same point when the second train catches up to the first, they will have traveled the same distance. Let t equal the number of hours the first train travels. Since the second train leaves two hours later, it will have traveled for two hours less. Therefore, $t - 2$ equals the number of hours the second train travels (see Figure 17-3).

Figure 17-3 Working through Example 4.

$$r \quad \times \quad t \quad = \quad d$$

first train $\boxed{40\text{ mph}} \times \boxed{t\text{ hr}} = \boxed{40t\text{ mi}}$

second train $\boxed{50\text{ mph}} \times \boxed{(t-2)\text{ hr}} = \boxed{50(t-2)\text{ mi}}$

The trains travel the same distance, so

$$50(t-2) = 40t$$
$$50t - 100 = 40t$$
$$10t = 100$$
$$t = 10$$

Therefore, it will be 10 hours after the first train leaves before the second train catches the first train. *But this is not the question!* The question asks, "At what time will the second train catch up to the first train?" The second train will catch up to the first train ten hours after 11:00 a.m., which is 9:00 p.m. The second question asks, "At that time, how far has each train traveled?" Replace t into either distance category and evaluate.

$$40t = 40(10) = 400 \quad \text{or} \quad 50(t-2) = 50(10-2) = 400$$

Therefore, each train has traveled 400 miles. Notice that the answer to the algebra is not always the answer to the question.

Example 5: Susan's boat can go 9 mph in still water. She can go 44 miles downstream in a river in the same time as it would take her to go 28 miles upstream. What is the speed of the river? Figure 17-4 shows this situation.

Figure 17-4 A current event, of sorts.

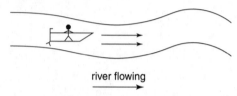

river flowing

When Susan goes in the same direction as the river, her speed increases by the speed of the river. When she goes against the river, upstream, her speed decreases by the speed of the river.

Let
$$r = \text{speed of the river}$$
$$9 + r = \text{Susan's speed downstream}$$
$$9 - r = \text{Susan's speed upstream}$$

Organize the information in a chart (see Figure 17-5).

Figure 17-5 A chart for the problem in Example 5.

	r	$\times$	t	$=$	d
downstream	$(9 + r)$ mph	$\times$	T_1	$=$	44 mi
upstream	$(9 - r)$ mph	$\times$	T_2	$=$	28 mi

Since $(r)(t) = d$, then $t = \frac{d}{r}$. So

$$T_1 = \frac{44}{9+r}, \quad T_2 = \frac{28}{9-r}$$

In the problem, the key phrase is "in the same time." So set the times T_1 and T_2 equal.

$$\frac{44}{9+r}=\frac{28}{9-r}$$
$$44(9-r)=28(9+r)$$
$$396-44r=252+28r$$
$$-72r=-144$$
$$r=2$$

Therefore, the river's speed is 2 mph. The check is left to you.

Work

The main formula for work problems is

$$\underset{\text{(rate of work)}}{(r)} \quad \underset{\text{times}}{\times} \quad \underset{\text{(amount of time worked)}}{(t)} \quad = \quad \underset{\text{work accomplished}}{a}$$

For example, if a person can do a job in four hours, then his or her rate of work is

$$\frac{1}{4}\,\frac{\text{job}}{\text{hr}}$$

A person's rate of work is the reciprocal of how long it takes to accomplish the job. Work problems are most easily done in terms of rate of work.

Example 6: George can mow a lawn in three hours. Joyce can do it in two hours. If they work together, how long will it take?

George needs three hours to do the job. Therefore,

$$\text{George's rate of work} = \frac{1}{3}\,\frac{\text{job}}{\text{hr}}$$

Joyce needs two hours to do the job. Therefore,

$$\text{Joyce's rate of work} = \frac{1}{2}\,\frac{\text{job}}{\text{hr}}$$

Let t = the time it takes to mow the lawn together and fill in the chart (see Figure 17-6).

Figure 17-6 A chart for the problem in Example 6.

	r	$\times$	t	$=$	a
George	$\frac{1}{3}\,\frac{\text{job}}{\text{hr}}$	$\times$	t hr	$=$	$\frac{1}{3}t$ job
Joyce	$\frac{1}{2}\,\frac{\text{job}}{\text{hr}}$	$\times$	t hr	$=$	$\frac{1}{2}t$ job

Together, they accomplish one job. Therefore,

$$\tfrac{1}{3}t + \tfrac{1}{2}t = 1$$
$$6\left(\tfrac{1}{3}t + \tfrac{1}{2}t\right) = 6(1)$$
$$2t + 3t = 6$$
$$5t = 6$$
$$t = \tfrac{6}{5}$$

Together, they will finish the job in $\tfrac{6}{5}$ (or $1\tfrac{1}{5}$) hours, which is the same as one hour and twelve minutes. The check is left to you.

Example 7: It takes Angela three hours longer to paint a fence than it takes Gary. When they work together, it takes them two hours to paint the fence. How long would it take each of them to paint the fence alone?

Let x = the number of hours it takes Gary to do the job alone.

$$\frac{1}{x}\frac{\text{job}}{\text{hr}} = \text{Gary's rate of work}$$

Let $x + 3$ = the number of hours it takes Angela to do the job alone (Angela takes three hours longer than Gary does).

$$\frac{1}{x+3}\frac{\text{job}}{\text{hr}} = \text{Angela's rate of work}$$

They complete the job in two hours, so fill in the chart as shown in Figure 17-7.

Figure 17-7   A chart for the problem in Example 7.

	r	$\times$	t	$=$	a
Gary	$\frac{1}{x}\frac{\text{job}}{\text{hr}}$	$\times$	2 hr	$=$	$2\left(\frac{1}{x}\right)$ job
Angela	$\frac{1}{x+3}\frac{\text{job}}{\text{hr}}$	$\times$	2 hr	$=$	$2\left(\frac{1}{x+3}\right)$ job

Together they finish one job:

$$2\left(\frac{1}{x}\right) + 2\left(\frac{1}{x+3}\right) = 1$$

$$\frac{2}{x} + \frac{2}{x+3} = 1$$

$$\cancel{x}(x+3)\cdot\frac{2}{\cancel{x}} + x\left(\cancel{x+3}\right)\cdot\frac{2}{\cancel{x+3}} = x(x+3)(1)$$

$$2x + 6 + 2x = x^2 + 3x$$

$$0 = x^2 - x - 6$$

$$0 = (x-3)(x+2)$$

Therefore, $x = 3$ or $x = -2$

Since x represents a length of time, $x = -2$ has no meaning in this problem. It takes Gary three hours and Angela six hours to paint the fence alone. The check is left to you.

Arithmetic/Geometric Series

Example 8: A grocery store display of soup cans has sixteen rows with each row having one less can than the row below it. If the bottom row has twenty-eight cans, how many cans are in the display?

This is the sum of an arithmetic sequence with sixteen terms. The sixteenth term is 28, and the common difference is 1. In order to use the formula

$$S_n = \frac{n}{2}(a_1 + a_n)$$

a_1 must be found. To do this, use the formula

$$a_n = a_1 + (n-1)d \text{ where } a_n = 28, n = 16, d = 1$$

$$28 = a_1 + (16-1)(1)$$

$$28 = a_1 + 15$$

$$13 = a_1$$

$$S_{16} = \frac{16}{2}(13 + 28) = 328$$

There are 328 cans in the display. Another approach could be to think of this as the sum of an arithmetic sequence with sixteen terms, the first of which is $a_1 = 16$ and whose common difference is $d = -1$. Then the direct use of the equation

$$S_n = \frac{n}{2}[2a_1 + (n-1)d]$$

yields the answer.

$$S_{16} = \frac{16}{2} [2(28) + (16 - 1)(-1)]$$
$$= 8(56 - 15)$$
$$= 8(41)$$
$$= 328$$

Example 9: A ball is dropped from a table that is twenty-four inches high. The ball always rebounds three fourths of the distance fallen. Approximately how far will the ball have traveled when it finally comes to rest?

Figure 17-8 shows this situation.

Figure 17-8 Follow the bouncing ball.

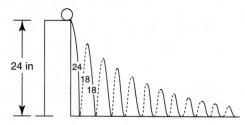

Notice that this problem actually involves two infinite geometric series. One series involves the ball falling, while the other series involves the ball rebounding.

Falling, $a_1 = 24, r = \frac{3}{4}$

Rebounding, $a_1 = 24(\frac{3}{4}) = 18, r = \frac{3}{4}$

Use the formula for an infinite geometric series with $-1 < r < 1$.

$$S = \frac{a_1}{1 - r}$$

$$S_{\text{falling}} = \frac{24}{1 - \frac{3}{4}} = \frac{24}{\frac{1}{4}} = 96$$

$$S_{\text{rebounding}} = \frac{18}{1 - \frac{3}{4}} = \frac{18}{\frac{1}{4}} = 72$$

$$S_{\text{falling}} + S_{\text{rebounding}} = 168$$

The ball will travel approximately 168 inches before it finally comes to rest.

For additional review of word problems, refer to *CliffsQuickReview Algebra I*.

Chapter Checkout

1. Sandy has $2,000 to invest. She invests part of it at 9% annual interest and the rest at 13%. At the end of a year, she has a total of $2,232. How much did she invest at each rate?

2. Bill wants to mix $5.20/lb cashews with $7.50/lb pecans to make 12 pounds of a mixture that will sell for $6.00/lb. How many pounds of cashews will be in the mixture?

3. A plane leaves City A at 9 a.m. heading due east at 530 mph, toward City B, which is 2,000 miles away. At the same exact time, a plane leaves City B, heading due west at 450 mph. At what time (to the nearest minute) will the two planes pass each other?

4. Bill can paint Mrs. Jones' fence in 4 hours. Sue can do it in 3. How long would it take Bill and Sue working together to complete the job?

Answers: 1. $700 @ 9%, $1300 @ 13%   2. 7.83 lbs.   3. 11:02 a.m.
4. $1\frac{5}{7}$ hrs. or 1 hr., 43 min.

CQR REVIEW

Use this CQR Review to practice what you've learned in this book. After you work through the review questions, you're well on your way to achieving your goal of understanding the basic concepts and strategies of Algebra II.

Chapter 1

1. Solve for x: $3x - 15 \geq 30$

   a. $x \leq 5$
   b. $\left| x \right| x \geq 5$
   c. $\left| x \right| x > 15$
   d. $\left| x \right| x \geq 15$

2. A parking lot is shaped like a trapezoid with area 152 sq. ft. One of its bases is 14 ft. long, and its height is 8 ft. Find the length of the other base.

Chapter 2

3. A line passes through points (–6, 4) and (6, 16). One equation for that line is given by:

   a. $-6x = 4y + b$
   b. $16y = 4x$
   c. $4x - 6y = 10$
   d. $x + 10 = y$

Chapter 3

4. Solve this system of equations by graphing:

$$\begin{cases} 2x + 4y = -22 \\ 5y - 6x = -2 \end{cases}$$

   a. $x = 3, y = -4$
   b. $x = 3, y = 4$
   c. $x = -3, y = -4$
   d. $x = -4, y = -3$

Chapter 4

5. Solve the following equation by using Cramer's rule:

$$\begin{cases} y + 4 = 5z \\ 7z + 15 - x = 28 \\ 4z + 11 = x + 3y \end{cases}$$

  a. $x = 1, y = 6, z = 2$
  b. $x = 2, y = 5, z = 7$
  c. $x = 3, y = 7, z = 5$
  d. $x = 8, y = 4, z = 6$

6. Solve this system for x, y, and z:

$$\begin{cases} 2x + 4y = 6 - 3z \\ x - 3y - 2z = -7 \\ x - 2y = -5 - z \end{cases}$$

  a. $x = 0, y = -2, z = 1$
  b. $x = -1, y = 2, z = 0$
  c. $x = -1, y = -2, z = 0$
  d. $x = 1, y = 0, z = 2$

Chapter 5

7. Find the difference:

$$(6x^4 - 3x^3 + 7x^2 - 5x) - (6x^4 - 3x^3 + 5x - 3)$$

  a. $7x^2 - 10x + 3$
  b. $7x^2 - 3$
  c. $-6x^3 + 7x^2 + 3$
  d. $-6x^3 + 7x^2 - 3$

8. Divide the following by using synthetic division. Express the remainder (if any) as a rational number.

$$\frac{x^4 - 3x^2 - 5x + 20}{x + 5}$$

  a. $x^2 - 5x + 22 - \dfrac{115}{x + 5}$
  b. $x^2 - 5x + 22 - 115$

c. $x^3 - 5x^2 + 22x - 115 + \dfrac{595}{x+5}$

d. $x^4 - 5x^3 + 22x^2 - 115 + \dfrac{595}{x+5}$

Chapter 6

9. Factor completely:

$$6x^2 - x - 15.$$

10. Factor completely:

$$12y^2 - 28y + 16.$$

Chapter 7

11. Simplify:

$$\frac{30x^2 + 55x + 15}{14x^2 - 76x + 30} \div \frac{6x^2 + 11x + 3}{21x^2 - 114x + 45}$$

  a. 6.5

  b. 7

  c. 7.5

  d. 8

12. According to Hooke's Law, the force needed to stretch a spring is proportional to the amount the spring is stretched. If 50 pounds causes a spring to stretch five inches, how much will the spring stretch when a force of 120 pounds is applied? (Hint: This is a direct variation problem.)

Chapter 8

13. If $f(x) = x^2 + 7x + 5$, find the following:

  a. $f(7)$

  b. $f(-3)$

  c. $f(0)$

14. If $f(x) = 3x - 7$, find $f^{-1}(x)$.

Chapter 9

15. Find all the zeros for the function:

$$f(x) = 3x^3 + x^2 - 12x - 4$$

   a. $\frac{1}{2}, \frac{1}{3}, 2$
   b. $\frac{-1}{3}, -2, 2$
   c. $-1, 0, 2$
   d. $\frac{2}{3}, -2, 2$

Chapter 10

16. Solve for b:

$$\sqrt{11b + 3} - 2b = 0$$

17. Simplify:

$$\left(2 - 3\sqrt{-98}\right) + \left(4\sqrt{-18}\right)$$

Chapter 11

18. Solve:

$$x^2 - 9x + 18 = 0$$

19. Solve:

$$4x^2 + 9x + 2 = 0$$

Chapter 12

20. Find the center and radius of the circle with the following equation:

$$x^2 + 4x + y^2 - 6y + 5 = 0$$

21. Find the focus, directrix, vertex, and axis of symmetry for the following parabola:

$$y^2 = 4x + 4y - 16$$

Chapter 13

22. Solve for a and b:

$$\begin{cases} ab = 2 \\ a^2 + b^2 = 4 \end{cases}$$

23. Solve for x and y:

$$\begin{cases} \dfrac{1}{x} = \dfrac{1}{y} + \dfrac{1}{4} \\ x \cdot y = 8 \end{cases}$$

Chapter 14

24. Solve and round the answer to the hundreths place:

$$4^{2x} = 17$$

   a. 0.99
   b. 1.00
   c. 1.01
   d. 1.02

Chapter 15

25. Find the fifteenth term of the arithmetic sequence:

$$1.00, 1.25, 1.50, 1.75, \ldots$$

   a. 4.25
   b. 4.50
   c. 4.75
   d. 5.00

Chapter 16

26. How many different 6-digit license tags can be printed if the first digit cannot be zero, and no digit may be used a second time?

27. How many permutations are there of the letters u, v, w, x, y, z?

Chapter 17

28. Phil takes 2 hours to paint 500 fence pickets, and Frank takes 3 hours to paint 450 fence pickets. How long will they take, working together, to paint 1000 fence pickets?

29. How many fluid ounces of distilled water must be added to 100 fluid ounces of a 15 percent sugar solution to make a solution that is 10 percent sugar?

Answers: 1. d, **2.** 24 feet, **3.** d, **4.** c, **5.** a, **6.** b, **7.** a, **8.** c, **9.** $(2x + 3)$ $(3x - 5)$ **10.** $(4y - 4)(3y - 4)$ or $4(y - 1)(3y - 4)$ **11.** c, **12.** 12 inches **13. a.** 103, **b.** –7, **c.** 5 **14.** $\frac{x+7}{3}$ **15.** b, **16.** 3 **17.** $2 - 9i\sqrt{2}$ **18.** 3, 6 **19.** $-\frac{1}{4}, -2$ **20.** $(-2,3); 2\sqrt{2}$ or $\sqrt{8}$ **21.** focus: $(4,2)$; directrix: $x = 2$; vertex: $(3, 2)$; axis of symmetry: $y = 2$ **22.** $\left(\sqrt{2},\sqrt{2}\right)\left(-\sqrt{2},-\sqrt{2}\right)$ **23.** $(2,4)$; $(-4,-2)$ **24.** d, **25.** b, **26.** 136,080 **27.** 720 **28.** 2.5 hours **29.** 50 fluid ounces

CQR RESOURCE CENTER

CQR Resource Center offers the best resources available in print and online to help you study and review the core concepts of Algebra II. You can find additional resources, plus study tips and tools to help test your knowledge, at www.cliffsnotes.com.

Books

This CliffsQuickReview book is one of many great books about Algebra II. If you want some additional resources, check out these other publications:

Ace's Exambusters Algebra 2/Trig. Study Cards, edited by Elizabeth R. Burchard, is a 384-page paperback collection of study cards that has proven helpful to many in studying for Regents and other standardized exams. Ace Academics Inc. $10.95.

High School Algebra Tutor, by James Ogden, starts with the rules and gives you many different problems and shows you how to solve them. *Algebra Tutor* does not go into depth on problems, but it is a great help as a reference book. Research & Education Association. $16.95.

Algebra the Easy Way, by Douglas D. Downing, is easy to work with and goes into full detail. A large number of problems to solve and their detailed solutions provide ample practice. Barrons Educational Series Inc. $12.95.

Algebra 2 : An Integrated Approach, by Roland E. Larson et al, is an excellent, if pricey, comprehensive volume that is useful for those planning to take math placement tests. At least one student reported that his confidence in his mathematical abilities was restored by this book. D C Heath & Co. $67.92.

Algebra Unplugged, by Kenn Amdahl and Jim Loats, reveals "secrets" of algebra that might make your math life much easier. It's nice, easy reading, with just the right mixture of humor and facts. Clearwater Publishing Co. $14.95.

Wiley also has three Web sites that you can visit to read about all the books we publish:

■ www.cliffsnotes.com

■ www.dummies.com

■ www.wiley.com

Internet

Visit the following Web sites for more information about Algebra II. There is a wealth of information out there, offering everything from help with homework to challenges, frequently asked questions (FAQs), and algebraic humor (and you didn't think math could be funny).

Stroh Math Page — www.homestead.com/stroh/mathpage.html— The page provides free algebra help, tutorials, practice problems, and live contact for specific assistance with Algebra I and II problems. Maintained by a math teacher and his students, the Stroh Math Page is a wonderful resource. Along with the help, you'll find links to many other math sites and math humor.

Mrs. Glosser's Math Goodies—www.mathgoodies.com—This is a free educational Web site featuring interactive math lessons, homework help, worksheets, puzzles, message boards, and more. It offers over 400 pages of free math activities and resources for students, teachers, and parents. You can join this growing community by subscribing to the free newsletter.

Ask Dr. Math—http://forum.swarthmore.edu/dr.math/dr-math. html—A forum at Swarthmore College, this site gives access to archives of previously asked math questions, as well as current FAQs. If you still have not found what you are looking for, then there's the link allowing you to ask Dr. Math his/herself.

Mathematica Demos—http://library.wolfram.com/demos/— This is a commercial site, where many different math solutions can be demonstrated. Although the site's owners want to sell you their program, Mathematica, the demos are available and can be viewed with a free downloadable viewer. Intersecting Functions and Systems of Equations are but two of the classroom demos that should be of interest to Algebra II students.

Mishawaka (IN) High School Math—www.mishawaka.k12.in.us/ mhs_files/dept/math/math.htm—This is a marvelous source for interactive math teaching and learning tools, as well as links to other great math Web sites and Math Lessons that are Fun (Rice University).

Purplemath Algebra Modules/Lessons —www.purplemath.com/ modules/—Looking for practical algebra lessons? These modules give practical tips, hints, and examples, and point out common mistakes. They are cross-referenced to each other, and a few of the modules contain "quizlets" to help you check your understanding. Included lessons for Algebra II students include completing the square; solving quadratics; finding a vertex; domain and range; and many more.

Next time you're on the Internet, don't forget to drop by www. cliffsnotes.com. We created an online Resource Center that you can use today, tomorrow, and beyond.

GLOSSARY

arithmetic sequence a sequence in which, starting with the second term, each term is found by adding the same value, known as the common difference, to the previous term.

arithmetic series the sum of the terms of an arithmetic sequence with a definite number of terms.

asymptote lines dashed lines on a graph representing the limits of values where a rational function or hyperbola is defined; a graph may approach its asymptotes, but will never reach them.

axis of symmetry (of an ellipse) either of the two axes intersecting at its center; the longer is the major axis, the shorter, the minor one.

axis of symmetry (of a parabola) the line that passes through the vertex and focus.

binomial an expression containing two terms separated by a + or − sign.

center the point in a circle from which all points are equidistant; in an ellipse, the midpoint of the segment joining the two foci.

circle a conic section; the set of all points in a plane equidistant from one point.

combination similar to a permutation, but when the order is not important. The combination of 8 objects taken 3 at a time would be $C(8,3)$ or $_8C_3$.

common difference can be found by taking any term in a sequence and subtracting its preceding term. See *arithmetic sequence.*

common logarithm understood to be base 10 when the base of a logarithm is not written. See *logarithm.*

common ratio found by taking any term in a sequence and dividing it by its preceding term. See *geometric sequence.*

completely factored incapable of being further simplified by division.

completing the square a technique for solving quadratic equations.

complex conjugates two binomials with the same two terms, but opposite signs, which represent the sum or difference of an imaginary number and a real number. For example, $a + bi$ and $a − bi$. See *conjugate.*

complex fraction a fraction containing one or more additional fractions (in the numerator, denominator, or both).

complex number any expression that is a sum of a pure imaginary number and a real number, usually in the form $a + bi$.

composite function a function in which the variable name has been replaced by another function.

compound inequality a mathematical sentence with two inequality statements joined by "and" or "or."

conic section cross section formed by a plane slicing through a point-to-point pair of cones; see *circle, parabola, ellipse,* and *hyperbola.*

conjugate axis the axis that passes through the center of the hyperbola and is perpendicular to the transverse axis. See *hyperbola.*

conjugates two binomials with the same two terms but opposite signs between them. For example, $5x + 3$ and $5x - 3$.

constant of proportionality the multiplier of the independent variable in a variation relationship (usually represented by k). For example, $y = kx$.

coordinates of a point the pair of numbers in the form (x,y) designating the location of any point on a plane.

Cramer's rule method of solving systems of equations by using determinants.

dependent system second version of the same equation, whose graphs coincides with each other.

descending order the general practice of writing polynomials in more than one variable so that the exponents decrease from right to left. For example:

$$2x^3 - 5x^2 + 3x + 2$$

determinant a square array of numerals or variables between vertical lines. A determinant differs from a matrix in that it has a numeric value.

difference of cubes an expression in the form of

$$a^3 - b^3$$

difference of squares a special pattern that is the result of the product of conjugates. For example $x^2 - y^2$ is the product of conjugates $(x + y)(x - y)$, $x^2 - 36 = (x + 6)(x - 6)$, etc.

direct variation "y varies directly as x" means that as x gets larger, y also gets larger.

directrix the line from which the set of points in a parabola are equidistant. See *parabola.*

dividend in a division problem, the number being divided into. See *quotient.*

divisor in a division problem, the number being divided by. See *quotient.*

domain set of all the x-values (first number of each ordered pair) in a relation.

ellipse a conic section; the set of points in a plane such that the sum of the distances from two given points in that plane stays constant. Each of those two points is called a focus point. The line passing through the foci is the major axis; its endpoints (on the ellipse) are its major intercepts. The line crossing the ellipse perpendicular to the major axis through the vertex is the minor axis. Its endpoints are at the minor intercepts.

equation a statement that says two mathematical expressions are equal.

exponential equation an equation in which the variable appears as an exponent.

exponential function any function defined by

$$y = b^x$$

extraneous solution solution that does not make the original equation true. Extraneous solutions are most likely to appear in equations that have been raised to a power or multiplied by a variable term to solve.

factor (n.) a number that is multiplied by another number to make a product. The factors of 6, for example, are 2 and 3, as well as 1 and 6.

factor, to (v.) to divide a polynomial by a constant or variable common to all of its terms. To divide. To rewrite a polynomial as a product of polynomials or polynomials and monomials.

factorial a way of expressing a natural number multiplied by all its preceding natural numbers. 4! is read "4 factorial" and means $(4)(3)(2)(1) = 24$.

first degree equation another name for a linear equation. See *linear equation*.

focus the point from which the set of points in a conic section are equidistant. In a circle, the focus is called the center. See *parabola, hyperbola,* and *ellipse.*

formula an algebraic equation that describes a rule, relationship, fact, principle, rule, etc. $I = PRT$, for example, is the formula for finding simple interest.

function a relation in which none of the domain values is repeated.

GCF (greatest common factor) the largest expression that can be factored (divided perfectly) out of another expression. For $3x^2 + 6x + 12$, the GCF is 3, yielding $3(x^2 + 2x + 4)$.

general term nth term of a sequence; a term of some order to be determined.

geometric sequence a sequence in which each term is found by multiplying the same value times the previous term. Taking any term in a geometric sequence and dividing it by its preceding term yields the common ratio.

geometric series the sum of the terms in a geometric sequence.

graph a pictorial display of solutions to mathematical equations. Also the point associated with an ordered pair.

greatest common factor see *GCF.*

hyperbola a conic section. The set of all points in a plane such that the absolute values of the difference of the distances between two given points stays constant; the two given points are the foci, and the midpoint of the segment joining the foci is the center. The transverse axis runs along the direction the hyperbola opens in. The conjugate axis passes through the center of the hyperbola and is perpendicular to the conjugate axis. The points of intersection of the hyperbola and the transverse axis are the vertices.

identity function $y = x$, or $f(x) = x$ since for each replacement, the result is identical to x.

imaginary value i represents $\sqrt{-1}$, which is an expression with no real value.

inconsistent system a system of non-intersecting equations. Their solution is the null set.

index in a radical expression ($\sqrt[n]{a}$), the n, which is an integer greater than 1. If a radical expression has no index, the index is assumed to be 2. See *radical expression*.

inequality a mathematical sentence using a relational symbol other than the equal sign (=).

inverse function a function in which the x and y variables have been switched; represented by $f^{-1}(x)$. No domain element appears twice.

inverse relation the set of ordered pairs created when the ordered pairs of the original relation are reversed.

inverse variation "y varies inversely as x" means that as x gets larger, y gets smaller, and as x gets smaller, y gets larger.

like radical expressions radical expressions with identical index and radicand. See *radical expression*.

linear equation an equation with one variable whose exponent is 1. The graph of a linear equation is a straight line.

linear inequality a linear sentence not containing an equal sign (=).

logarithm exponent expressing the power to which a fixed number (the base) must be raised in order to produce a given number. Abbreviated as *log*. It is usually computed to the base 10 (common logs, where the base is not written), or to the base e (known as natural logs and abbreviated ln); the purpose is to shorten mathematical calculations.

logarithmic equation an equation that involves the logarithm of an expression containing a variable.

logarithmic function a function of the form

$$y = \log_b x$$

where $x > 0$, $b > 0$, and $b \neq 1$

major axis the line passing through the foci of an ellipse, having its endpoints on the ellipse. See *ellipse*.

major intercepts the points where the major axis of an ellipse touch the curve itself. See ellipse.

matrix (pl. matrices) a rectangular array of numerals or variables that can be used to represent systems of equations.

minor axis see ellipse.

minor intercepts see ellipse.

monomial a single term expression, not containing separate parts separated by + or − signs. For example: 5, x, $3a$, $4x^2y^2$.

multiplication principle for events a principle used to determine how many different ways a particular event can occur. For example, if one event can occur in p different ways and another in q different ways and p and q are independent events, then together they can occur in pq different ways.

natural logarithm a term that represents log base e (also $\log_e$), which is written as ln. See *logarithm*.

ordered pair represented as (x,y). The x-value always comes first, separated from the y-value by a comma. See *coordinates of a point*.

origin the point $(0,0)$ where x-axis and y-axis intersect.

parabola a conic section. The set of points in a plane that are the same distance from a given point and a given line in that plane. The given line is called the *directrix*, and the given point is called the *focus*.

Pascal's triangle a graphical representation of binomial expansion, named after the French mathematician Blaise Pascal.

permutation the arrangement of objects in a certain order. For example, 8 objects arranged 3 at a time would be $P(8,3)$ or $_8P_3$.

point-slope form (of a non-vertical line) takes the following form, where $(x - x_1)$ = difference in x- coordinates, and $(y - y_1)$ = difference in y- coordinates; m is the slope.

$$y - y_1 = m(x - x_1)$$

polynomial an expression consisting of terms separated by some combination of + signs, − signs, or both.

polynomial function any function of the form

$$P(x) + a_0x^n + a_1x^{n-1} + a_2x^{n-1} + \cdots$$
$$+ a_{n-1}x + a_n$$

where the coefficients $a_0, a_1, a_2, ..., a_n$ are real numbers, and n is a whole number.

proportion an equation stating that 2 rational expressions are equal.

pure imaginary number any product of a real number and i. For example: $3i$, $5i$, etc. See *imaginary value*.

quadrants the four regions defined by the intersection of the x- and y-axes and designated by Roman numerals. Beginning in the top right and proceeding in a counterclockwise direction, quadrant I is the top right; quadrant II the top left; quadrant III the bottom left, and quadrant IV the bottom right.

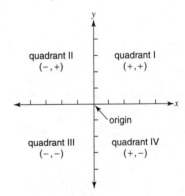

quadratic equation any equation in the following form:

$$ax^2 + bx + c = 0 \qquad (a \neq 0)$$

quadratic form any equation of the following form; such equations may be solved by quadratic formula:

$$ax^{2n} + bx^n + c = 0$$

quadratic formula a formula that may be used to solve any and all quadratic equations in standard quadratic form:

$$x = \frac{-b \pm \sqrt{b^2 - 4ac}}{2a}$$

quotient the answer to a division problem. In $10 \div 5 = 2$, 10 is the dividend, 5 is the divisor, and 2 is the quotient.

radical the bracket also known as the "square root" sign (if its index is 2).

radical equation an equation in which the variable is under a radical sign.

radical expression the name given the following:

$$\sqrt[n]{a}$$

The bracket is known as the radical sign; *a* is the radicand, and *n* is the index. If no *n* appears on the radical sign, the index is assumed to be 2. The above is read as "the *n*th root of *a*."

radicand the number under the radical. See *radical expression*.

radius the distance from the center of a circle to any point on the circle.

range set of all the *y*-values (second number of each ordered pair) in a relation.

rational equation an equation involving rational expressions.

rational expression the quotient of two polynomials, usually expressed as a fraction. The denominator may never be zero.

rational function If $f(x)$ is a rational expression, then $y = f(x)$ is a rational function.

rationalizing the denominator a process used to remove radicals from denominators of rational expressions. To rationalize a denominator, multiply by the conjugate of the denominator over itself.

relation a set of ordered pairs.

sequence an ordered list of numbers.

slope intercept form $y = mx + b$, where *x* and *y* are the coordinates of a point on the graph of the line, *m* is the line's slope, and *b* is some constant.

slope of a line the line's rise over its run (or its change in *y* divided by its change in *x*) as the graph of the line moves to the right. A line that descends as it moves right has a negative slope; a horizontal line has a slope of 0; a vertical line's slope is undefined.

square trinomial the expression produced by squaring a binomial:

$$(x + y)^2 = x^2 + 2xy + y^2$$
$$(x - y)^2 = x^2 - 2xy + y^2$$

standard form of a line the standard form for the equation of a line is

$$Ax + By = C$$

where *A*, *B*, and *C* are integers and *A* is positive.

sum of cubes an expression in the following form:

$$a^3 + b^3$$

summation notation uses the Greek letter sigma (Σ) to indicate the sum of a finite number of terms in a sequence. The lower number, or index of summation, is the value where summation starts. The upper number is the upper limit of summation:

$$\sum_{d=3}^{8}(3d-7)$$

synthetic division a shortcut for dividing a polynomial by a binomial of the form $x - a$; only coefficients are used.

term any number in a sequence or piece of a polynomial separated by a + or − sign.

transverse axis the line along the direction the hyperbola opens in passing through its vertices. See *hyperbola*.

trinomial an expression containing three terms separated by + or − signs.

varies directly as one quantity increases or decreases, so does another quantity. See *direct variation*.

varies inversely as one quantity increases another decreases, and vice versa. See *inverse variation*.

vertex (of hyperbola) either of the two points of intersection of the hyperbola and the transverse axis. See *hyperbola*.

vertex (of parabola) the midpoint of the perpendicular segment from the focus to the directrix.

vertical line test a test for functions. If a vertical line passes through more than one point on a graph, then a domain point has been repeated, and the graphed relation is not a function.

x-axis the horizontal axis; all points with a *y*-coordinate of 0.

x-coordinate the number to the left of the comma in an ordered pair.

x-intercept the point at which a graph crosses the *x*-axis.

y-axis the vertical axis; all points with an *x*-coordinate of 0.

y-coordinate the number to the right of the comma in an ordered pair.

y-intercept the point at which a graph crosses the *y*-axis.

zero of a function any value for the variable that will produce a solution of 0.

Index

continued

vertex of a hyperbola, 186–189, 259
vertex of a parabola, 174, 176–179, 259
vertical line test for functions, 122–123, 259

W

Web sites, 3, 250, 251, 252
word problems
    arithmetic/geometric series, 241–242
    compound interest, 235
    general strategy for, 233–234
    mixture, 236
    motion, 236–239
    simple interest, 234
    work, 239–241
work problems, 239–241

X

x-axis, 22, 259
x-coordinate, 23, 259
x-intercept, 29, 259
x-numerator determinant, 49

Y

y-axis, 22, 259
y-coordinate, 23, 259
y-intercept, 29, 259
y-numerator determinant, 49

Z

zero product rule, 93–95, 154
zero values
    asymptote lines and, 113–115
    equations of lines and, 29
    LCD, 107
    product rule for radicals and, 144
    rational expression denominator and, 97
    rational function denominator and, 113
    rational zero theorem, 135–136
    slope of lines, 26
    as y-coordinate of x-intercept, 29
    zero product rule, 93–95, 154
    zeros of a function, 135, 259
zeros of a function, 135, 138, 259